UNDERSTANDING ENERGY

UNDERSTANDING ENERGY

Energy, Entropy and Thermodynamics for Everyman

REFERENCE

R. Stephen Berry

Department of Chemistry and The James Franck Institute
The University of Chicago

World Scientific
Singapore • New Jersey • London • Hong Kong

Published by

World Scientific Publishing Co. Pte. Ltd.

P O Box 128, Farrer Road, Singapore 9128

USA office: Suite 1B, 1060 Main Street, River Edge, NJ 07661

UK office: 73 Lynton Mead, Totteridge, London N20 8DH

Library of Congress Cataloging-in-Publication Data

Berry, R. Stephen, 1931–
    Understanding energy : Energy, entropy, and thermodynamics for
everyman / R. Stephen Berry.
       p.   cm.
    ISBN 981020342X. -- ISBN 9810206798 (pbk.)
    1. Force and energy. 2. Entropy. 3. Thermodynamics. I. Title.
QC73.B43 1991
531'.6--dc20   91-19564
                                                         CIP

Printed in Singapore by Loi Printing Pte. Ltd.

SIGNS AND SYMBOLS

a acceleration

a_0 Bohr radius

A   availability

$\propto$ coefficient of thermal expansion

c speed of wave fronts, i.e. speed of light in empty space

C_p constant pressure

C_V constant volume

d distance

d   infinitesimal increment

Δ small but macroscopic increment

e electronic charge

E   energy

ε effectiveness

$\mathscr{E}$ change in potential

η efficiency

F   force

$\mathscr{F}$ Helmholtz free energy

$\mathbb{F}$ Faraday, or electric charge/mole of electrons

g constant (9.80 m/sec)

g   gram

G   universal constant

g degeneracy

$\mathscr{g}$ Gibbs free energy

h	height
h	Planck's constant
H	enthalpy
i	species
j	quantum number
J	quantum integer
k	constant of proportionality
k_B	Boltzmann's constant
κ	compressibility
κ	heat conductance
KE	kinetic energy
ℓ	angular momentum quantum number
L	length
λ	wavelength
m	mass
m	meters
M	number of molecules
μ	chemical potential
n	number of moles (counting Avogardro's number at a time)
n	number of little pieces of anything
n	principal quantum number
N	number of particles
N_A	Avogadro's number
ν	frequency of a wave
p	pressure

p	momentum
P	external pressure on a system
$P(E_j)$	probability of finding a system with energy E_j
PE	potential energy
q	Coulomb attraction
Q	heat gained by a system
r	radii
R	distance between objects
$\mathbb{R}$	gas constant
R_∞	Rydberg constant for an infinitely massive nucleus
R_H	Rydberg constant for the real hydrogen atom
s	number of components
S	entropy
t	time
T	temperature
v	velocity
V	volume
$\mathbb{V}$	volume of 1 mole
w	work
w	waste factor
W	work function
$\mathbb{W}$	number of microstates
$\propto$	is proportional to
$\equiv$	is equivalent to, or is defined as
$\approx$	approximately equal to

NATURAL CONSTANTS AND CONVERSION FACTORS

Quantity	Symbol	Value
Speed of light	c	3.0×10^8 m/sec
Mass of the electron	m_e	9.12×10^{-31} kg
Mass of the proton	m_p	1.67×10^{-27} kg
Charge of the electron	e	-1.60×10^{-19} coulomb
Bohr radius	a_o	0.529×10^{-10} m
Rydberg constant	R_∞	1.097×10^7 m or 2.18×10^{-18} joule
Boltzmann's constant	k_B	1.38×10^{-23} joule/deg.K
Gas constant	$\mathbb{R}$	8.31 joule/mole deg.K or 2.0 cal/mole deg.K
Avogadro's number	N_A	6.02×10^{23} things
Faraday	$\mathbb{F}$	9.65×10^4 coulomb/mole
1 electron-volt (eV) is equivalent to or to		1.602×10^{-19} joule 23,000 calories/mole

FOREWORD

This text is intended for two audiences, neither of them scientists or engineers. One is composed of undergraduate humanists and social scientists who want a taste of a fundamental physical science, particularly one that bears on their own lives. The other audience consists of professionals in fields outside science and engineering who now find themselves needing to understand the arcane jargon that scientists use when they talk about energy and other natural resources. The manuscript began as notes for undergraduate courses in Physical Science at The University of Chicago, which I first gave in 1968 and revisited many times subsequently. During the 1960s, '70s and '80s, I frequently found myself talking with attorneys, public officials and concerned citizens (some with, some without capital letters) with strong desires to learn more of the basic science underlying the technical aspects of issues that we were facing.

In the style of The University of Chicago, I wanted not to give courses limited to current issues, but to focus on the intellectual principles that we must call upon to deal with them and with other issues in the future. Consequently the main goal of this work is communicating a set of clean and precise physical concepts: the abstract ideas underpinning macroscopic and microscopic mechanics, the manner in which we use generalizing concepts to reduce unwieldy complexity in describing physical systems, and the natural laws that seem to us to govern the relationships among these abstract ideas. The most important of these for my two audiences are the laws governing energy and the efficient use of natural resources. This means that expounding the concepts and laws of thermodynamics is the ultimate aim of the text. All the other sciences — the mechanics, the kinetic theory, the atomic and molecular theory, the theory of radiation and quanta — are here to clarify and provide background for the thermodynamics. One can develop that magnificent subject as a self-contained logical structure without mechanics, quantum theory, kinetic theory or statistics, but only at the risk of limiting the lay reader's exposure to the breadth and power of its concepts. In teaching physical science, my own most successful approaches have always involved exploring energy, heat, temperature and related abstractions first from the primitive levels of mechanics, both the classical version of Isaac Newton and its modern quantum counterpart, and then from the view of what these imply in complex macroscopic systems. This is my approach here, to give just enough background to expose the very rich concept of energy in several contexts before it plays its part as a thermodynamic variable. Then, in thermodynamics, I, like many others, have found it very helpful to complement the macroscopic approach to entropy with a discussion of the conceptually simpler microscopic and statistical picture.

Thermodynamics is indeed a magnificent science. For the sheer economy of its generalizations of experience, it stands alone among all human intellectual creations. It is not a fashionable science; the trend of twentieth-century physics has been reductionist, that is, "explanation", to mean interpretation at an ever-more-microscopic level. This approach will always be part of science, of course. But there is another way of looking at the natural world. This way is in terms of relations we believe are universal, among quantities we can observe directly or compute from observables, that describe the macroscopic world. The discovery of these relations, which we call scientific laws, is a remarkable triumph of human intelligence; even more remarkable is the invention of the concepts and variables which satisfy those relations. Therein is the mystery and the fascination of thermodynamics.

CONTENTS

PREFACE

This text deals with the concept of energy and its intimately related concepts from thermodynamics. The goal is the development of these concepts, in terms of their relation to underlying mechanical phenomena, their meaning at the submicroscopic level of atoms and molecules, and their bearing on familiar processes on the human scale. Thermodynamics teaches us about some of Nature's constraints, about the relation of time and reversibility, and about the significance of natural constraints for an evolving technological society.

The material divides into two sections. The first twelve chapters deal with energy from the viewpoint of microscopics and mechanics. They introduce the concepts of forms of energy and of "degrees of freedom" or places where energy can be stored, and of the conservation of energy. Laying the basis for the First Law of Thermodynamics at several levels is, in a sense, the intellectual goal of the first part of the text.

The second part treats energy from the more global, generalized view embodied in the First and Second Laws of Thermodynamics. Here we examine the generalities that follow from the conservation of energy, and the way energy and matter flow with time. It develops the concepts of thermodynamic state functions such as entropy and thermodynamic potential, and shows how these reveal the efficiency of machines and processes, and the limits of our capabilities to convert heat energy into useful work.

1. THE BASICS, I: FORCE AND WORK

FORCE

The Laws of Mechanics, as developed by Copernicus, Kepler and Newton, have come down to us expressed in terms of a number of concepts that take on very precise and quantitative meanings in the context of scientific discussion. Those we shall be using are the concepts of **mass, force, acceleration, momentum, velocity, position**, and **energy**. (There are others, but we need not discuss them here.) We use the same words frequently in common vocabulary, but often without intending to give them the precise meanings they take on in science. In the following discussion, we shall need to use their precise, scientific significance, so let us approach them by examining the laws that give them meaning.

The Law of Inertia, Newton's First Law, can be stated to say that the momentum of an isolated object remains unchanged. **Momentum** here means specifically (**mass**) × (**velocity**) or, abbreviated, mv. Mass, position and time are the three physical quantities we leave as primitive intuitive notions, defined only in terms of arbitrary reference standards and measurements with balances, rulers and clocks. Velocity means the rate at which position changes with time: miles *per* hour or meters *per* second or in general, distance *per* unit of time. "Per" implies "divided by"; a velocity is, in effect, a distance divided by a time. This is the first of several instances in which we can identify a common word such as "per" with a particular mathematical operation or relation, in this case division. **Acceleration** is the rate at which velocity changes with time: (miles per hour) *per* hour, or (meters per second) *per* second. Like positions on a line running from $-\infty$ to $+\infty$ (read "∞" as "infinity") velocities and accelerations may be positive or negative. Positive velocity means the moving object is moving toward larger positive distances, and negative velocity, toward more negative distances. Positive acceleration means either an object is moving to more positive directions and is speeding up, or that it is moving toward more negative directions and is slowing down. The crux, in either case, is that the velocity is becoming more positive or, equivalently, less negative. Now, as an exercise, give a comparable interpretation of what negative acceleration means.

Newton's Second Law of Motion deals with acceleration, not with velocity. This Law states that the acceleration experienced by an object is directly proportional to the **force** on that object, and that the mass of the object is exactly the proportionality factor. The equation

$$ma = F \qquad\qquad (1)$$

is exactly equivalent to a sentence saying that the mass of the object, multiplied by the rate at which the velocity changes with time, is equal to the force on the object. The Law says nothing about what kind of force F might represent; this force could be a push or a pull, a shake or a jiggle, or no force at all. The force could be constant or varying in time, uniform throughout space or a function of the position of the object. If we were to measure m and watch the acceleration of the particle, keeping track with rulers and a clock, we would actually be mapping the force on the particle as a function of space and time. We say we would be measuring or mapping the *field of force*.

One might be tempted to charge that the Second Law of Motion only *defines* force to be a quantity that assures that the equation $F = ma$ is satisfied. In one sense, this is exactly what the Law does. Historically, the Law transformed a diffuse notion into a precise and quantitative idea by requiring "force" to mean that quantity with dimensions of (mass) × (distance)/(time)^2, equal to (mass) × (acceleration). If every pair of particles interacted through its own particular sort of force, the Second Law of Motion would be little more than a tautology, and might never have been proposed; surely it would be of very little use. But we do discern some very important regularities in nature. Within our experience, our universe exhibits only a very few types of force, so that it becomes natural to attribute independent existence to each kind of force field. We think of the *force* field as the observable consequence of mapping the way a test object would accelerate if it could be put anywhere in the vicinity of some source, such as a magnet, a charged metal plate or a massive object. The source has the potential to establish a force field with any test object, so we attribute to the source itself the notion of a *potential* field, a concept we shall study shortly, without reference to any particular test object. The Second Law of Motion tells us how the motion of the test object is related to the force field.

Whether the force field has an independent reality or is an abstract construction of the human mind is a question for metaphysics, not for science. The answers to this question have come back variously as "no", "yes" and, from the logical positivists, "it is a meaningless question". The great import for science lies in the fact that we can use the Second Law of Motion to connect various observations and make predictions of real, observable processes. We can use it to discover the forms of force fields and to recognize the apparent universality of various kinds of forces. We can use the Second Law to predict the future positions and velocities of objects from a knowledge of their positions and velocities at a single time, once we know the force field in which the objects move.

The gravitational force field is one of the simplest. For any two objects, with masses m_1 and m_2, the gravitational force is *always attractive*, and its strength is inversely proportional to the square of the distance R between the objects. The constant of proportionality is the product of the masses, m_1 and m_2, multiplied by a universal constant G, whose value (6.672×10^{-11} meters^3 kg^{-1}sec^{-2}) we must infer from observation. The force law for gravitation has the form

$$\text{Force (gravitational)} = -(m_1 m_2 / R^2)G \; ; \qquad (2)$$

G has the value just given if the masses are in kilograms, the distances in meters, the time in seconds and the force in kilograms × meters × seconds^{-2}. The negative sign means that the force tends to make R smaller − − i.e. to draw the particles together.

If the mass of one of the objects, say m_2, is much larger than the mass m_1 of the other, then the lighter particle number 1 does virtually all the moving as a consequence of the force. In this case, the force law describes the acceleration a_1 of particle 1 due to the presence of heavy particle 2, which remains almost fixed in space. We shall not derive this relation here, but we can write it as a mathematical sentence or equation,

$$m_1 a_1 \approx m_1 (m_2 G) R^2 \qquad\qquad (3a)$$

or

$$a_1 \approx m_2 G / R^2 \qquad\qquad (3b)$$

in which we see that the acceleration of particle 1 depends on its distance R from particle 2, on the universal constant G and on the mass of particle 2 but *not* on its own mass. This is quite an accurate description of a person or a ball close to the earth's surface, interacting with the earth's gravitational field of force.

The transformation from the general equations **(1)** and **(2)** to the special equation **(3a)** involves a detailed consideration of how to deal with the masses of two interacting objects in the expression "mass × acceleration". The mass m_1 on the left side of Equation **(3a)** enters in a very different way than m_1 on the right side of the same Equation **(3a)**. The mass m_1 on the left is associated with the *general* First Newtonian Law of Motion; this is the so-called *inertial* mass. The masses m_1 and m_2 on the right appear in the very special context of one kind of force law, the gravitational force; these are called the *gravitational* masses. It is a general belief among scientists that these two kinds of mass are equivalent. This belief in the **Principle of Equivalence** has been subjected to some very stringent tests, so that there is strong experimental evidence that the Principle is at least approximately true. We can never hope, in the absence of a theory far more general than any we now have, to *prove* the Principle of Equivalence; we can only subject it to ever more demanding tests and use it as though we were convinced of its validity.

The **Law of Gravitation** has a close parallel in **Coulomb's Law** which governs the interaction of electrostatic charges (stationary electric charges). Coulomb's law states that two charged objects interact with each other, exerting a force of interaction inversely proportional to the square of the distance between the objects, just as in the Law of Gravitation. However the electrostatic force between charged objects may be repulsive or attractive. Charges may be positive or negative; if both particles carry charges of the same sign, both positive or both negative, the force is repulsive, and if the charges have opposite signs, one positive and one negative, the force is attractive like the force of gravitation. In the precise language of an equation, if the objects 1 and 2 carry charges q_1 and q_2, the separation between objects is R, and no other medium or objects are present to add other contributions, then the force

$$F = q_1 q_2 / R^2 \ . \qquad\qquad (4)$$

(There may be a proportionality constant multiplying the right-hand side, depending on the choice of units.) Note that if q_1 and q_2 have opposite signs, F is negative (attractive) like gravitation, while F is positive if q_1 and q_2 have the same sign.

If we suppose that mass m_2 is larger than mass m_1, again only particle 1 is moved significantly by the influence of the force, so that we can write the Second Law of Motion once more in terms of a_1, the acceleration of particle 1, as

$$m_1 a_1 \approx q_1 q_2 / R^2 \qquad\qquad (5)$$

The similarity of Eq. **(5)** to the gravitational force of Eq. **(3a)** is striking indeed. The difference is the nature of the numerators on the right sides of the equations. In Eq. **(3)**, m_1 appears on both sides of the equation and therefore cancels; in Eq. **(5)**, this is not the case.

Another force of interest to us is the force associated with two test objects separated by a variable distance R and connected by an ideal spring, as shown in Fig. 1. This force tends to decrease R if we stretch the spring and to increase R if we squeeze the spring. At one value of the distance R, the spring has no tendency to stretch or shrink. This is the **equilibrium point** of the spring; we denote that distance by R_e, the distance between the test objects on the ends of the spring at which there is no force tending to move them.

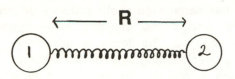

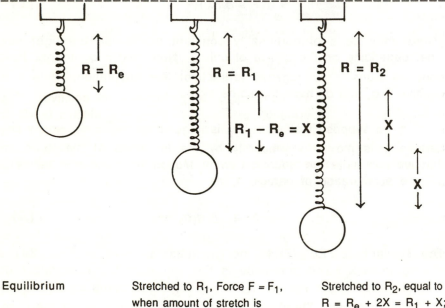

Figure 1a. Two particles, 1 and 2, with masses m_1 and m_2, held together by a compressible, extendable spring.

Equilibrium

Stretched to R_1, Force $F = F_1$, when amount of stretch is $X = R_1 - R_e$, or distance $R = R_e + X$

Stretched to R_2, equal to $R = R_e + 2X = R_1 + X$; Force $F_2 = 2 F_1$

Figure 1b: The Ideal Spring or Harmonic Oscillator

There is a simple model of an *ideal spring*. This special, simple model has one special property: not only does the force act to return the spring to its equilibrium point, but the restoring force is exactly proportional to the displacement of the spring from its equilibrium position.   Suppose the spring is stretched so that the objects at the ends are a distance R_1 apart, instead of R_e, with $R_1 > R_e$.   Then there is a force F trying to shrink the spring toward R_e, as shown in Fig. 1b.   The distance $X = R_1 - R_e$ is called the **displacement** from equilibrium. If the displacement is doubled to 2X, so that the distance becomes $R = R_e + 2X$, then the special property of the ideal spring or *harmonic oscillator* is that the restoring force doubles, from F_1 to $2F_1$.

More generally, the force restoring a harmonic oscillator to its equilibrium position can be expressed in a simple equation, in terms of the displacement of the oscillator.   We may write

$$F(\text{harmonic oscillator}) = -kX \ . \qquad (6)$$

We assume that k, the constant of proportionality, is positive. The negative sign in the equation assures that the force always tends to make X smaller in absolute magnitude; X is greater than zero if the spring is stretched and the force is positive, acting to reduce X. If the oscillator is squeezed, then the new position R_3 is less than R_e, and $X = R_3 - R_e$, the displacement, is *negative*. Then if X is negative, the force $-kX$ is positive and F acts to *increase* X algebraically toward zero.   Thus, whether R is less than or greater than R_e, whether X is negative or positive, the force expressed in Eq. **(6)** tends to return X to zero and R to R_e.

Equations **(3)**, **(5)** and **(6)** describe three different examples of kinds of forces. There are others, such as the forces exerted by magnets, and the forces that hold atomic nuclei together, and the **net effective forces** that result from the total interactions between composite structures – – the forces between molecules, for example, or the forces between charged particles in a liquid. We shall not need to consider these other forces here.

WORK

Now consider what happens if one tries to move an object subjected to a force. As a most mundane example, take the object as your own body and the force as the gravitational interaction between the earth and your body. To step up onto the first stair in a flight of stairs, you must act against gravity; a certain amount of effort must be spent. Going up ten stairs requires somewhat more effort; going up ten flights of stairs requires considerably more. The process of climbing stairs gives us an intuitive, physiological, sense of what is involved in moving *against a force*. Climbing up stairs is a process of *doing work*. Climbing five stairs requires only half as much work as climbing ten stairs because the force of gravitational attraction is essentially the same at the fifth stair or the tenth stair as it is at the bottom of the staircase. The force of the earth's gravity is essentially constant, independent of altitude, so long as one remains very close to the earth's surface.* When the force (any force) has this property of being constant in space, independent of position, then the amount of work required to move against that force increases directly with the distance the object is moved. Moving up a thousand stairs requires a thousand times as much work as moving up one step. We say that the work increases *in direct proportion* to the distance moved, and, as well, in direct proportion to the constant force, more specifically, in direct proportion to their product. There is no extra constant of proportionality so that

$$\text{Work} = \text{Force} \times \text{Distance}$$

$$w = F \times d \; . \qquad\qquad (7)$$

----

*   Suppose the object 2 is essentially a sphere with radius R_2, and that particle 1 is a point. Particle 1 may be at a distance R from the center of particle 2, greater than or equal to ($\geq$) R_2. Suppose there is some maximum R of interest to us, which we call R_2'. Think of R_2 as the earth's radius and R_2' as the height of the highest mountain measured from the earth's center. We let $h = R_2' - R_2$, the range of heights or distances we care about. This is much less than R_2 itself. Then in this special situation, the forces $m_1 m_2 G/(R_2')^2$ and $m_1 m_2 G/R_2^2$ are almost identical:

$$\frac{m_1 m_2 G}{(R_2')^2} = \frac{m_1 m_2 G}{R_2^2 + 2R_2 h + h^2}$$

$$= \frac{-m_1 m_2 G}{R_2^2} \times \frac{1}{[1 + 2(h/R_2) + (h/R_2)^2]}$$

but $(h/R_2) \ll 1$, so $(h/R_2)^2$ is even smaller and the factor in brackets is very nearly 1, making the force very nearly constant, with the value

$$m_1 m_2 \, G/R_2^2 .$$

When we use this picture to describe the earth and things near it, object 2 is the earth with radius R_2 and mass m_2. In this case, we define, for the sake of abbreviation, the gravitational acceleration $g \equiv m_2 G/R_2^2$, where m_2 is the mass of the earth and R_2 is its radius.

If we deal with objects such as spacecraft, that we want to lift large distances, then we can no longer assume that the gravitational force is constant. Clearly, the Law of Gravitation says that the force decreases as the distance R, of the object from the center of the earth, increases varying inversely with R, that is, as $1/R^2$.

It is not difficult to calculate the work required to move a rocket to an altitude of, say, 10,000 kilometers. One does this by treating the rocket like an object moving up a very long succession of short staircases. Suppose m is the mass of the rocket. The gravitational force on the first staircase is F_1, which is equal to mG/R_1^2; R_1 is the distance from the staircase to the center of the earth. Suppose the staircase is 3 meters high. Then the work required to lift the rocket up the first staircase is $3F_1$. The second staircase is $R_1 + 3$ meters from the center of the earth, so the force of gravity at the second staircase is $mG/(R_1 + 3)^2$. Let us call this force F_2. Suppose the second staircase is also 3 meters high. Then the work to lift the rocket up the second staircase is $3F_2$. Because $(R_1 + 3)^2$ is a bit bigger than R_1^2, the force F_2 is a bit smaller than the force F_1, and the work $3F_2$ is a bit less than the work $3F_1$. Going up the next 3 meter staircase requires a bit less work, $3F_3$, where F_3 is $mG/(R_1 + 6)^2$ or $mG/(R_2 + 3)^2$. The differences between $3F_1$ and $3F_2$ or between $3F_2$ and $3F_3$ are small. However when we get up to an altitude of 1000 or 10,000 kilometers (1,000,000 or 10,000,000 meters, respectively), the situation is quite different. The force on the millionth 3 meter staircase, for example is $F_{one\ million} = mG/(R_1 + 3 \times 1,000,000)^2$. The radius of the earth is R_2; it is slightly over 6,000,000 meters. The force F_1 is $mG/(6,000,000)^2$, but the force at the millionth 3 meter staircase is $F_{one\ million} = mG/(9,000,000)^2$, considerably less than F_1. In fact,

$$\frac{F_{one\ million}}{F_1} = \frac{mG/(9,000,000)^2}{mG/(6,000,000)^2} = \frac{6^2}{9^2} = \frac{4}{9}$$

In other words, the force at the millionth staircase is only four-ninths of the force at the first staircase at the launching pad on the earth's surface. As an exercise, compute the ratio of the force at the top staircase to the force F_1, if the top staircase is 10,000 kilometers above the earth.

The work required to lift the rocket up each little 3 meter staircase is the force at that staircase (which we persuaded ourselves that we could assume constant over a range of only 3 meters), times the distance of 3 meters. The *total work* is the sum of the amounts of work required at each staircase. But the value of the nearly-constant force must be recalculated at the beginning of each new staircase. In an explicit statement, we can write the sentence: Total work = $3F_1 + 3F_2 + 3F_3 + \ldots 3F_{last}$. It is easy to see that if the work were the same for each staircase, i.e. if the forces were the same at all the staircases, we could write the total work as total work = force × total distance, as we did with a single staircase.

The point of this exercise has been twofold. First, we have seen the relation between force and work. Second, we have seen that we can express work in an *explicit and quantifiable way*, even for forces that vary with position, by using the trick of dividing the total distance into small steps, so small that the force is almost the same at both ends of the step. This means that mechanical work takes on a precise meaning, as well as its intuitive, visceral meaning of what one does when one acts against a force.

We have used an example of work thus far that involves someone or something working against gravity. The concept of work is more universal than this, and can be used in reference to movement against any kind of force. In general, for any constant force F, the work *w* done to move something against that force for a distance *d* is,

$$\text{work } w = F \times d \ .$$

For a force that varies with distance, we chop the distance *d* into n little pieces $X_1, X_2, \ldots X_n$, so that

$$d = X_1 + X_2 + \ldots + X_n \ .$$

Then we evaluate the force at each of these pieces; at X_1, the force is F_1; at X_{17}, the force is F_{17}, and so on. In this case, the work can be expressed as a sum of little contributions, just as it was with the rocket:

$$w = X_1 F_1 + X_2 F_2 + \ldots + X_n F_n. \tag{8}$$

[Query: How can we find the work we try to represent by this sum if we decide that the force near the beginning of the first interval beginning at X_1 is, after all, rather different from the force near the end of that same first interval?]

Now suppose we have lifted a 100 kilogram ball up to the top of a 3 meter staircase. We have done a fair bit of work; is it possible to regain this work?   The answer is that, in an ideal, purely mechanical system, it is.   We could let the weight of the ball compress a spring, or turn a wheel, or perform in any number of ways to recover our work, simply by letting gravity pull the ball back down the 3 meters to the earth.   We need not get useful work from the descent of the ball, but we may if we choose.   We have only to put up a suitable device to harness the falling motion.   By lifting the ball, we have given it the capacity to do work.   This stored capacity to do work, and its generalizations, will be one of the central concepts of our investigation.

Problems

1. How long would it take for a 1 lb weight to fall the 1454 feet (443.2 meters) from the top of the Sears Tower to the ground? (Assume that the friction of the air can be neglected.) What would the time of fall and final velocity be if the weight were 10 lbs instead of 1 lb? What is the potential energy of the 1 lb weight at the top of the building, in joules (1 joule = 1 newton-meter)?

2. How long would it take for an object weighing 0.1 kilogram with a charge of +0.1 coulomb to fall one meter toward a fixed spot charged with −0.1 coulomb, located two meters away? The law of electrostatic forces between two point charges has the form F(newtons) = Const $\times q_1 q_2 / R^2$, where q_1 and q_2 are the charges (in coulombs) R is the distance between the charges (in meters) and the constant is 9×10^9. What would be the final velocity of the weight? What would be the time of fall and final velocity if the weight were 1 kilogram instead of 0.1 kilogram? The unit of force, the newton, is 1 kilogram-meter-(seconds)$^{-2}$.

3. Why do the time of fall and final velocity of the moving body depend on the mass of the body in the case of electrostatic force but do not depend on the mass of the moving body if the force is gravitational?

4. The radius of Mars is approximately half the radius of the earth. The mass of Mars is approximately one-tenth the mass of the earth. Show that the gravitational acceleration constant g on Mars is approximately two-fifths as much as g on earth.

NOTE: g = 32 ft/sec^2 or 9.8 m/sec^2

   $G = 6.67 \times 10^{-8}$ cm^3/gm sec^2 or 6.67×10^{-11} m^3/kg sec^2

   R_{earth} = 6370 km or 6.37×10^6 m

   $g = G/R^2_{earth} = 1.05 \times 10^{-19}$ m/kg sec^2

   $m_{earth} = 5.9 \times 10^{24}$ kg

2. THE BASICS, II: KINETIC AND POTENTIAL ENERGY

Suppose an object is first at rest and then is subjected to a force; for example, a baseball rests in a pitcher's hand and is then accelerated by the catapulting force of the throwing arm. The force of the pitcher's arm does *work* on the ball by changing its velocity from zero to some finite value, v_f. The *change* in velocity, final velocity minus initial velocity, is simply v_f in this example, because the initial velocity is zero. This change is accompanied by a change in the **momentum** of the ball. The momentum of any object is its **mass** multiplied by its **velocity**. The momentum of the ball changes from zero to a value equal to the final velocity v_f, multiplied by its mass, m. The ball in motion has the capacity to do work on some other object, by transmitting a force to that other object, provided the other object has any velocity other than v_f in the direction of motion of the ball. (If the ball and the other object both had velocity v_f in the same direction they would never meet.) This capacity to do work, in the form of motion of the ball, is called **kinetic energy**. The kinetic energy of the ball is determined by the observer who considers herself stationary, and sees the ball moving with velocity v_f. The kinetic energy is equal to the maximum amount of work that the ball could do if it were to strike its target and be left standing still, relative to the observer. The value of the kinetic energy, KE, of an object with mass m and velocity v is given by

$$KE = \tfrac{1}{2}mv^2 \ . \qquad\qquad (9)$$

This expression is general and will appear again and again, when we consider the energy of moving atoms, molecules and massive objects.

Suppose an object is initially located at some place from which it can be accelerated. For example, the object could be a ball in a pitcher's hand, a rock at the top of a hill, or a weight at the end of a compressed spring. Each of these is in a position where it is subjected to a force (and perhaps also a restraining force, just to hold the object while we observe it). The objects may have no kinetic energy when we first observe them, this would mean that they would have initial velocities of zero. However, if we remove the restraining forces, then the pitcher's arm, the earth's gravitation or the restoring force of the spring will accelerate the object from its initial position and bring its *kinetic* energy from its initial value (zero if the initial velocity was zero) to a value $mv^2/2$, where v is its velocity at the moment we measure the kinetic energy.

In their initial positions, the objects had the potential or capacity to gain kinetic energy or to do work by being accelerated. We refer to this *stored capacity* of a mechanical system, such as an object in a force field, as the **potential energy** of the system. The particular value that the potential energy takes on depends on the kind of force involved. The value of the potential energy of a simple mechanical system can always be obtained by determining the amount of work required to prepare the system. The **total energy** of a system is its kinetic energy plus its potential energy. Because *total energy* is *conserved*, and in a simple mechanical system, there are no ways to dissipate energy (simple mechanical systems have no friction, for example), we can always recover any energy put into a mechanical system as work. The *potential energy* gained by lifting a rock from the bottom of a hill to the top is exactly the amount of work required to lift the rock up the hill. The potential energy stored in the compressed ideal spring is equal to the amount of work required to compress the spring.

Work and energy are thus very intimately related. They have the same units, of (force) × (distance). However work and energy are *not* the same. Energy is a property of a system: it may be kinetic energy of motion, or potential energy associated with position, but in either case, energy is a quantity or a property of the system in its environment. Work, on the other hand, is characteristic of a process, or of the transformation of a system. Work is what one performs, spends or receives, when one changes the energy of a simple, mechanical system. We may speak of an amount of energy being used as work, but we do not speak of a system "containing work". Rather, we speak of a system as containing an amount of potential energy, which means that the system contains the capacity to perform an equivalent amount of work.

In a simple, ideal mechanical system, we emphasize again, there is no way to use, transform or dissipate energy except as work. In the more complex systems with which we shall be concerned later, energy may be put into work, but also may be dissipated in ways we would not classify as work, ways we call heat. Before looking for the meaning of heat, it is very important to develop a clear idea of the meaning of potential and kinetic energy and of work in simple mechanical systems; then we can add the complexity of heat to our discussion.

Problems

1. A heavy fly weighs 1 gram. Estimate its average speed, in centimeters/second, and calculate its approximate, average kinetic energy in gram-centimeter^2/second^2.

2. How many kilograms does a bowling ball weigh, approximately? How fast does it travel when bowled by a skilled player in meters/second? What is its approximate kinetic energy in joules (kilogram-meter^2/second^2)?

3. What is the average momentum of a 4-minute miler? Estimate the runner's mass and calculate the average velocity during the run. Use units of kilograms, meters and seconds.

3. INTERCHANGE OF KINETIC AND POTENTIAL ENERGY

HYDROELECTRIC POWER

The potential energy of a system subjected only to gravity depends only on its position. Thus, the potential energy of an object in the earth's gravitational field depends only on the height of the object above the earth. The higher the object, the longer it can be accelerated, the greater will be its final velocity and the more is its capacity to do work.

Consider, for example, the water at the top of a dam. One liter of water (weighing 1000 grams, or 1 kilogram), dropped through the distance h from the top of the dam to the bottom of the dam, (that is, h is the height from the intake to the outfall) decreases its potential energy by an amount (mass) × (height) × (acceleration due to gravity) or mgh, which is (1 kg) × (gh) in this example. Suppose the dam is 100 meters high. The gravitational acceleration at the earth's surface, g, which we can suppose constant is 9.80 m/sec^2, which we may approximate as 10 m/sec^2.* Lifting one liter of water up 100 meters would cost approximately (1 kg) × (10^2 m) × (10 m/sec^2) or 10^3 kg m^2/sec^2, which is 10^3 joules per liter of water. This is stored as potential energy as long as the water is kept at that 100 meter level, and is also just the amount of energy that could be converted into work when we let the water fall to the bottom of the dam.

The maximum amount of work we can recover by letting that liter of water be pulled down over the dam is thus 10^3 joules. This energy could be extracted in many forms; often, it is taken out as

---

* In using this approximation, we make an error of 20 parts in 980 or approximately 2 parts in 100, i.e. 2%. If we can tolerate a 2% error – – which is certainly acceptable in the present context – – the approximation is more useful than the exact calculation because it is so much easier and quicker to use.

---

electrical energy, when the falling water is made to turn the vanes of a turbine connected to an electric generator. A large waterfall may have a flow of 500,000 or 5×10^5 liters/sec which would, in principle, allow us to withdraw energy at a rate of (5×10^5 liters/sec) $\times$ (10^3 joules/liter) or 5×10^8 joules/sec. The unit measure of the flow of energy in time, the energy transformed per second, is called **power**. The basic metric unit of power is the joule per second, which is called the **watt**. One thousand watts is of course a kilowatt, and a million watts is a megawatt. Hence our large hypothetical dam is capable of generating 5×10^5 kilowatts or 500 megawatts of power. This is approximately the capacity of Bonneville Dam in Washington state and is roughly a tenth the size of the largest hydroelectric power plants now in operation.

The process of hydroelectric power generation is particularly convenient because the energy to do the work of lifting the water comes to us on a rather regular basis from the sun. Water, by evaporating and diffusing into the sky, acts as the acceptor for solar energy; the energy to do the lifting work comes into the water as heat and light, and is transformed partially into potential energy when the water becomes a cloud. This stored energy is released when the water vapor condenses into drops and the drops fall to earth. We rarely try to use all the energy of falling rain to do work, but we do harness the last bit of fall of the water from high altitudes on the earth's surface down to sea level, in order to capture potential energy and do work.

Real hydroelectric plants can never operate as perfect energy conversion systems. Typical plants can convert 80-90% of the potential energy of their stored water into electrical energy. When we examine the conversion of heat energy to electrical energy, we shall see that the hydroelectric system performs very well in terms of the conversion of potential energy into work. However we shall also see that lifted water, the source of energy for a hydroelectric system, is a less effective medium for storing energy than a fuel such as gasoline or diesel oil, in the sense that the potential energy stored per unit mass or per unit volume of elevated water is much less than the potential energy in an equivalent mass or volume of a combustible fuel.

STABLE ORBITS AND STATES

Despite the way it looks to us, the earth's surface is rather smooth and therefore almost a surface of constant potential energy for any object on that surface. But there are bumps and hollows on the earth, and the potential energy of objects on this surface depends on location. The higher the point on the surface, the greater is the altitude h and therefore the greater is the potential energy mgh, for an object whose mass is m. Any object initially held at some high point to keep it from falling, when released, moves naturally from a point of high potential energy to a point of lower potential energy and, in so doing, gains kinetic energy by converting some of its potential energy to kinetic energy. This phenomenon is not confined to the neighborhood of the earth's surface. A comet far from the solar system is drawn toward the sun by the attractive force of gravitation. A planet in the gravitational field of the sun is attracted toward the sun. As these objects are drawn toward the sun, they accelerate, which is the same as saying that their velocities increase.

Occasionally, an object literally falls into the sun, captured by the sun's gravitational field. To suffer this calamity, the object's trajectory must pass within the boundary of the sun. More commonly, objects drawn toward the sun accelerate and miss the solar disc, passing *around* this source of attractive force. They then move away from the sun, decelerating as they leave. During the acceleration phase, the sun does work on the planet or comet, just as the earth accelerates a rolling stone. During the deceleration phase, as it moves outward, the planet, comet or stone regains its potential energy and loses kinetic energy.

Many comets appear only once. These comets are presumably one-time visitors that have enough energy initially to escape completely from the sun's gravitational field. In contrast, the planets and some comets have insufficient energy to escape the sun's attractive force, yet have enough energy to remain in stable periodic orbits around the sun. A ball rolling down one side of a frictionless valley and up the other side, then down and up the first side again, back and forth, is another example of an object in a stable trajectory. We speak of objects in such non-escaping, persistent trajectories as being in *stable bound orbits*. Objects that move so slowly or in such narrow orbits that they fall into the sun are in *unstable orbits*. (What happens to these objects *in* the sun is a completely different question, and lies outside this discussion.) Likewise, comets that wander in from far outside the solar system, go once around the sun and then escape are in *free orbits*.

It is convenient to think of the motion of an object in *any* force field as the motion of that object on an appropriate potential energy surface. For objects small compared with the size of the earth and close to a surface like that of the earth, where the gravitational force is almost constant, we can *approximate* the potential energy surface as a simple hill with the potential energy PE(h) as

$$PE(h) = mgh . \qquad \qquad (10)$$

This is an approximation, and is only valid when h is small compared with the size of the earth. When we wish to deal with larger distances, we must use the more general and accurate representation of the potential energy in a gravitational field,

$$PE(h) = - \frac{m_1 m_2 G}{(h + r_1 + r_2)} = \frac{m_1 m_2 G}{R} , \qquad (11)$$

where we assume objects 1 and 2 are spheres with radii r_1 and r_2, respectively, and R is the distance between their centers. This is essentially the same as the problem of moving the rocket off the earth, discussed in Chapter 1. In other words, the gravitational potential energy varies inversely with the separation of the objects. This is a consequence of the inverse-square form of the force law, as in Eq. **(2)**, so that the potential energy of two charged particles must also vary inversely with their distance. The approximation of

Eq. **(10)** reflects the situation in which h in Eq. **(11)** is tiny compared with either R_1 (the size of object 1) or R_2, which may be the radius of the earth.

Exercise: Express the gravitational acceleration in terms of the more basic quantities m_{earth}, R_{earth} and G.

Note that the zero of the potential energy PE(h) in Eq. **(11)** occurs when h goes to infinity, and all other values of PE(h) are less than PE(∞). However in the approximate Eq. **(10)**, the lowest value of PE(h) occurs if $h = 0$ and PE(h) is greater than zero when h is greater than zero. These two choices correspond to different selections for the distance at which we choose to say the potential energy is zero. One is zero at the surface of the larger object and the other is zero when the objects are infinitely far apart. We are really only concerned with changes in PE(h), not in the absolute value of PE(h), so that as long as we are consistent and maintain the same choice for the zero of PE(h), there is no problem. The choice of the zero of PE(h) is arbitrary, so we pick it as conveniently as possible for each situation.

Exercise: Show that two scales of potential energy differing by an additive constant yield the same *changes* in potential energy for the same process.

It is helpful to see how PE(h) looks graphically; Figures 2 and 3 show the two cases we have discussed so far. The graphs show, in essence, a cutaway section of a hill. Any unconstrained body placed on one of the hills will roll down, trading potential energy for kinetic energy. Note how any very tiny portion of Figure 3 looks like the straight line of Figure 2. We often describe a region of low potential energy, such as the region around $h = 0$ in Figure 3, as a *potential well*, using the natural analogy with a real well.

One other type of potential is important to us, the potential energy associated with the ideal spring or *harmonic oscillator*. The true gravitational potential grows less steep at large distances.

The constant-slope potential is just that, growing neither steeper nor flatter. The simplest potential that grows steeper with increasing separation is that of the harmonic oscillator. As a conceptual starting point, the harmonic oscillator is perhaps the most widely used model in physics and chemistry.

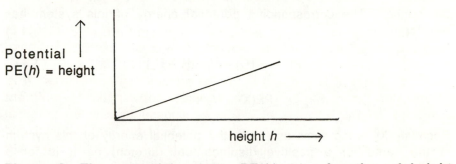

Figure 2. The potential energy PE(h) as a function of height h, when PE(h) increases linearly with h.

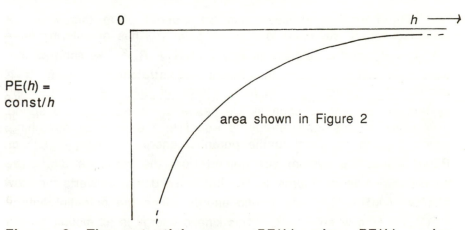

Figure 3. The potential energy PE(h), when PE(h) varies inversely with h.

Recall that the *restoring force* on an object attached to an ideal spring has the form (Eq. **(6)**)

$$\text{Force(spring)} = -k(R - R_e) = -kX \qquad (12)$$

where $X = R - R_e$, the displacement from the equilibrium position of the spring. The corresponding potential energy of this system has the form

$$PE(R - R_e) = \tfrac{1}{2} k(R - R_e)^2 \qquad (13)$$

or

$$PE(X) = \tfrac{1}{2} kX^2 \quad .$$

Because X^2 is the same as $(-X)^2$, the potential energy of this system is the same for a positive displacement (stretch) as it is for a negative displacement (compression) of an equal amount. The form of the potential energy is that of a parabola. The force and potential energy are shown in Figures 4a and 4b.

The harmonic oscillator can be thought of in this way. A weighted spring, stretched to a given point R_1, has a restoring force $-k(R_1 - R_e)$ and a potential energy $\tfrac{1}{2} k(R_1 - R_e)^2$. As soon as the weight is released from its (restrained) stretched position, it is accelerated toward R_e. The *total* energy of the idealized stretched system is constant as soon as we cease to do work on the spring. The value of this total energy is equal to the work of stretching, which, in turn, is equal to the potential energy $\tfrac{1}{2} k(R_1 - R_e)^2$ or $PE(R_1)$ that the system has gained from the work we did. The accelerated system returns to R_e, but in so doing, converts the total energy $\tfrac{1}{2}k(R - R_e)^2$ into kinetic energy, since the potential energy $PE(R_e)$ is zero when $R = R_e$. This kinetic energy is all motion "to the left" in Figure 4b, i.e. motion toward compression. The momentum of the moving weight is now capable of doing work to *compress* the spring. In fact the total kinetic energy of the weight, $\tfrac{1}{2}mv^2$, can now be put back into potential energy of *compression*. The precise amount of potential energy that the system can gain is equal to its total energy, $\frac{1}{2}k(R - R_e)^2$, and will be attained when $R = R_e - R_1$. At this point, the potential is

$$PE(R_e - R_1) = \frac{1}{2}k(R_e - R_1)^2$$

$$= \frac{1}{2}k(R_1 - R_e)^2 \qquad (14)$$

$$= PE(R_1 - R_e), \text{ the initial potential.}$$

Then the now-compressed spring will have come to a dead stop. At this instant, the system will start to expand, accelerate through R_e, slow up to R_1, stop and repeat the entire cycle. The weight will simply *bounce* between $R_e + R_1$ and $R_e - R_1$. Equivalently, we can think of the system as a particle sliding up and down the parabolic well between $R_e + R_1$ and $R_e - R_1$ in Figure 4b. This picture gives us a physical image of the motion of the system *on a potential surface*, in contrast to a mechanical model in the form of a spring. The notion of a potential surface will occur many other times during our discussions. In this example, the surface is just a simple parabolic curve; later we will examine richer cases.

One further point should be made concerning the potential surfaces and the motion of systems on them. The surface of Figure 4b rises to an infinite height. Any system on such a surface, which has only a finite amount of energy E, is constrained to remain between the points at which all the energy is converted into potential energy, that is, for which PE(R) = E. In other words, the potential $PE(R) = \frac{1}{2}k(R - R_e)^2$ keeps all systems *bound*, so that *all* orbits of the harmonic oscillator are stable.

By contrast, the potential shown in Figure 3, has a maximum value of zero, approached as h becomes very large. Any system whose total energy is greater than zero can and will escape from that potential well. Particles with energies less than zero are trapped in the well. Such a potential has both *bound* and *free*, or *stable* and *unstable* orbits. We shall see several other systems of this type, as well as systems that behave like harmonic oscillators in the sense of having only bound orbits.

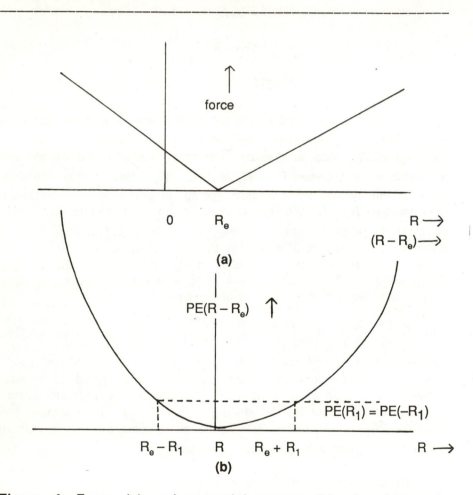

Figure 4. Force (a) and potential energy (b) of a harmonic oscillator, as a function of displacement from equilibrium.

PROBLEMS

1. A spring has a constant k of 0.5 newton/meter. About how far do you think you could stretch this spring? (That is, how many newtons of force can an average person exert?) Suppose this particular spring is harmonic until it has been stretched 40 cm (0.4 m), at which point the spring breaks. How much work must one do to break the spring?

2. In a heavy rain, the accumulation of water is 1 cm. The rain all falls from a cloud at an altitude of 1000 meters or 1 kilometer. Neglecting any friction due to the atmosphere, calculate the total amount of energy of the rain falling onto 1 square meter, when it reaches the earth.

3. Consider a smooth ball on a) a level, frictionless table top and b) a rough table top. Use these two cases as the basis of illustrations of stable, unstable and neutral equilibrium.

4. A typical American house requires about 22 million kilocalories or 2.2×10^{10} calories of heat from fuel each year, equivalent to about 9×10^{10} joules. This energy is usually supplied from chemical energy, by burning coal, oil or gas. Suppose you lived in a house near a river; would it be feasible to heat this house by converting waterpower into heat? You will have to construct your own hypothetical model of a river; assume you can build a relatively small dam, but not over 10 meters high, and that the river is not a very large one. (See p. 22) Assume that your own neighborhood river has a flow between 1000 liters/sec in the dry season and 10,000 liters/sec at its maximum of full flood. You can make a pond large enough to maintain a steady flow somewhere between these limits.

5. In the normal atmosphere, there are about 2.7×10^{19} molecules in each cubic centimeter. What is the *flux* of molecules – –

that is, the net number of molecules crossing a unit area, of 1 cm^2 in this case – – in a whole gale wind of 60 miles/hour, or 3000 cm/sec? What is the momentum of a single molecule with mass 1.8×10^{-22} g moving with this velocity? What is the *pressure* – – the force on a unit area, 1 cm^2 of a wall – – exerted by this wind? Use the fact that force is momentum transferred per second, and therefore is the momentum transferred to the wall by one molecule, times the number of molecules that strike the 1 cm^2 of wall in one second. What is the total force on a window 1 $meter^2$ in area, if the wind strikes it directly?

6. Estimate by a few bar graph approximations the minimum work (in units of liter atm) required to fill the 1 liter tire from a pressure of 1 atm to a pressure of 6 atm at 20°C. That is, estimate the area under the curve of p *vs* V from $p = 1$ to $p = 6$ atm. Start by calculating the volume that would be occupied by the air in the pumped up tire if that air were allowed to expand until its pressure was exactly 1 atm. This is the *total initial* volume; the initial pressure is of course 1 atm. The final volume is 1 liter. The final pressure is 6 atm. Now draw a curve of p *vs* V and estimate the area equivalent to the work of compression. What is the amount of work in *joules*? (One liter atm is 25.8 joules.)

4. WORK AND ENERGY RELATIONS FOR IDEAL GASES

The connections among work, potential and kinetic energy and force for simple mechanical systems can be extended to gases. In particular, we begin by supposing our gas to be composed of atomic particles that have no internal ways of storing energy. These particles have only *kinetic* energy.

Let us consider how this gas does work. Suppose the gas is contained in a chamber, one of whose walls is a sliding piston. The work done by the gas is determined by the *force against which* the piston is moved, multiplied by the distance through which the piston moves: work = (force against which motion is carried out) × distance. The questions we must ask now are these: how is the energy of this ideal gas, originally in the form of kinetic energy of the gas particles, converted into work? How is the gas affected by having done work? How do we describe the state of the gas, in order to answer the first two questions?

The **state** of a macroscopic system such as a box of gas is a concept that pervades much of physical, biological and social science. By "the state of a system", we mean a statement of the condition of the entire system that, in some sense, summarizes the average mechanical condition of the vast number of its individual atoms or molecules without specifying the motion of each one of the particles. It is unthinkable that the state of a gas would specify the position and velocity of each particle of a gas, just as it is unreasonable to expect a description of the economic state of a nation to describe the income and expenditures of each citizen. We require that a small number of summary variables specify the state of our system. It is not clear what the most appropriate variables should be for a social system, but we do know a great deal about the variables appropriate for describing the state of a physical system.

The variables that we have found particularly suitable to describe the state of a physical system are: its temperature, its pressure, its density, its mass, its volume, its energy, and its composition. One could naturally find others that are combinations of these variables, but the values of these additional variables could be derived from a knowledge of the original set.[*] The realization that new variables would be redundant (or at least contained in the old ones) leads us to question whether the temperature, pressure, energy, volume, mass, density and composition are all independent. The answer is "NO"; there *are* relations among these variables, and we need not specify them all. For a pure system consisting of a single substance, the composition is automatically specified; if we know that the system is all gas, all liquid or all solid – – i.e. in the form of a single *phase* – – then we need specify only *two* of the other variables temperature, pressure, volume, density or energy. All the others can be derived from an equation called the **Equation of State**, so called because it expresses the connections among the state variables. Each substance and each phase (solid, liquid or gas) has its own equation of state. However certain classes of systems, like all dilute gases, are well described by a single equation of state, independent of substance, if the equation is put into an appropriate set of units. Part of our task in this chapter will be to derive the equation of state for the ideal gas, to show how the state variables are related to each other and to the "microscopic" variables that describe individual atoms or molecules. We shall look first at the form taken by the expression for mechanical work, appropriate for describing gases or other systems of many atoms, and then see how the variables that arise naturally in that expression are connected by an equation of state. We shall, in the process, offer a simple derivation of the equation of state for an ideal gas, and thereby connect the mechanical variables of the atomic system to the variables describing the macroscopic properties of the complex system, variables that summarize the microscopic properties.

---

[*] This set of properties or variables is enough to describe the states of many, many systems, but some, such as systems subjected to magnetic or electric forces, or systems such as foams and aerosols, require more variables for their description. For example the strength of the magnetic field and the extent of magnetization of the system are important for systems subjected to magnetic forces; surface area and surface tension are important variables for foams and aerosols.

Pressure is the first in our list of summary variables. What is pressure? Its definition is

$$\text{pressure } p = \text{force per unit area} \quad . \qquad (15)$$

The force exerted by the atmosphere on a square meter of the earth's surface is equivalent to the force of 1.03×10^4 kilograms of mass acted upon by the earth's gravitation field of 9.80 m/sec^2. This is equivalent to 101,300 newtons. It is also equivalent to the force exerted by the earth's gravitational field (at the earth's surface) on a column of mercury, Hg, 760 millimeters high. Hence we say that the *pressure* of the earth's *atmosphere* at the earth's surface is 1 atmosphere or 1.013×10^5 newtons/m^2 or 760 mm Hg or 0.760 m Hg. (The unit of 10^5 newtons/meter^2, equal to 10^6 dynes/cm^2, is called the *bar*; the unit equivalent to the pressure due to a 1 mm column of mercury is called the torr, after the Italian experimenter Torricelli.)

If a gas pushes a piston against a force, then the work done by the gas is

$$\text{work} = (\text{resisting force}) \times (\text{distance piston is moved}) \qquad (16)$$

$$= \frac{\text{resisting force}}{\text{area of piston}} \times (\text{area of piston}) \times (\text{distance piston is moved})$$

The first parentheses of the second line of Eq. **(16)** enclose a quantity equal to the *pressure* on the piston from the outside, and the product of the second and third factors is just the change ΔV in *volume* V, of the system. This is illustrated in Figure 5. Hence the work done by a gas pushing a piston is

$$\text{work} = (\text{external pressure on the gas})$$
$$\times (\text{change in volume of the gas})$$

$$= (p)(\Delta V) \text{ or } p\Delta V \quad . \qquad (17a)$$

If the external pressure is constant, we can evaluate the work by a simple multiplication. If the external pressure changes as the piston moves, then we must evaluate the work in small steps and add the contribution from each step, as we did in computing work done against a force that varied with position:

$$\text{Work} = p_1(\Delta V)_1 + p_2(\Delta V)_2 + \ldots + p_n(\Delta V)_n . \qquad \textbf{(17b)}$$

Piston Area = A

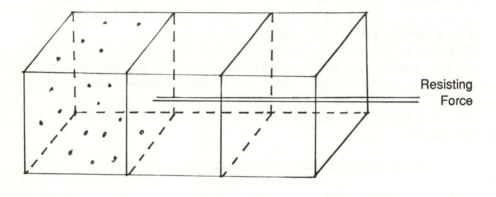

Resisting Force

Displacement
ΔX

Change of volume of gas = $A\Delta X$

Figure 5. Work done by a gas pushing a piston

The dimensions and units of work are the same as the dimensions of energy. Work is characteristic of a process, but energy characterizes a state, so work is associated with changes of energy. We may measure work and energy in *joules*, as we have thus far (1 joule is 1 kg meter^2/sec^2) or in *ergs* (1 erg is 1 g cm^2/sec^2, or 10^{-7} of a joule). We may also use other convenient measures: taking p in atmospheres and V in liters, work done by a gas can be evaluated in liter-atmospheres (1 liter-atmosphere is about 100 joules). We can also measure work and energy in the units customary for heat, such as calories, which we shall discuss later. (One thousand calories, or 1 kilocalorie (often called 1 $\underline{C}$alorie) is about 4 joules.)

Next, how is the energy that goes into work done by the gas related to the kinetic energy of the gas molecules? (Recall that we assume for the present that the gas particles have only kinetic energy.) To see this, let us suppose that the pressure exerted by the gas on the piston is essentially the same as the external pressure, so that the piston never has a chance to accelerate. We determine the pressure exerted by the gas particles colliding with the piston. To do this, recall that the *force* on a wall or a particle is equivalent to the momentum transferred to that wall or particle, per second.

$$\text{force} = (\text{mass}) \times (\text{change of velocity})/(\text{interval of time})$$

$$= m(v_{\text{final}} - v_{\text{initial}})/(t_{\text{final}} - t_{\text{initial}})$$

$$= m\,\Delta v/\Delta t$$

$$= (mv_{\text{final}} - mv_{\text{initial}})/(t_{\text{final}} - t_{\text{initial}})$$

But mv, (mass) $\times$ (velocity), is momentum. Hence

$$\text{force} = \text{change of momentum}/\Delta t \qquad \qquad (18)$$

Hence we can compute the pressure, that is, the force per unit area, if we can compute the momentum transferred to the wall of the piston per unit area, per unit time, typically per second. We do this by computing the momentum transferred per collision, and multiplying this by the number of collisions per second, as follows.

Suppose, in order to simplify our model (but without losing any of the essence of the model), that the molecules are all moving with the same absolute velocity $|v|$, along one or another direction parallel to the walls of the container,* and that at any instant, one-sixth of the molecules move toward each of the six faces of the chamber, which we now assume to be a simple box or parallelepiped.

Particles moving toward a particular wall have a momentum mv before they strike the wall; rebounding perfectly, their velocities change from v to $-v$, so they have a final momentum $-mv$. The momentum of a particle changes from $+mv$ to $-mv$ when a collision occurs; the difference, $2mv$, must be transferred to the walls by the collision, if momentum is to be conserved. Hence $2mv$ is the momentum transferred, per collision.

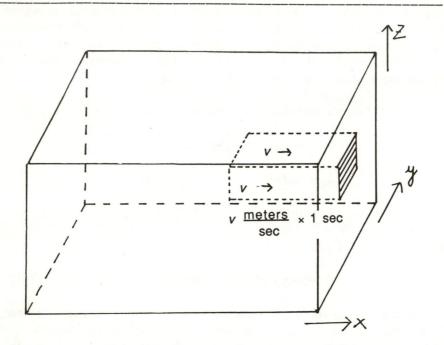

Figure 6. Gas molecules colliding with the walls of their container

* In Chapter 15, we shall see how this assumption can be replaced with the more plausible one, that the molecules may have any speed, but that the number of molecules with any given speed is, on average, unchanging with time. With this more general assumption, we shall reach the same conclusions as we reach in this chapter except that the quantity v^2 appearing in **(20)**, below, will be replaced by the *average* of v^2 for the gas molecules.

How many collisions do the molecules make with the walls in one second? Suppose there are N molecules in a big box whose volume is V, so that there are N/V molecules per unit volume in the box. Construct a hypothetical little box extending from one face of the big box, as in Figure 6. Suppose its area shared with a face of the box is 1 m^2. At any instant, one sixth of the molecules in the little box are moving along in the x direction toward the shaded 1 m^2 portion of the wall of the real box, as shown. A molecule moves a distance of (v m/sec) × (1 sec) = v m in one second. Hence, if the little box is made v m long, then exactly one-sixth of the molecules in the little box will collide with the real wall in one second. The number of collisions with the 1 m^2 bit of wall, in one second, is therefore

$$\text{\# of collisions/(m}^2 \text{ sec)} =$$

$$(N/V)(\text{molecules/m}^3) \times 1/6 \times (v \text{ m/sec}) =$$

$$1/6 \times (N/V)v \text{ collisions/(m}^2 \text{ sec)} . \qquad (19)$$

Therefore the total force on one m^2, or the total momentum transferred to 1 m^2 of wall in one second, which is exactly the pressure on the wall, is

$$2mv \text{ (kg m/sec) per collision} \times 1/6 \text{ N/V } v \text{ collisions/(m}^2 \text{ sec)}$$

or $\qquad\qquad$ $\frac{1}{3} mv^2$ N/V , so that we may write $\qquad$ **(20)**

$$p = \frac{1}{3} mv^2 \text{ N/V} .$$

Hence the *pressure* of a gas is directly proportional to the number N of particles in the container, inversely proportional to the volume V of the container, and directly proportional to $\frac{1}{3} mv^2$, which looks almost like the kinetic energy of a molecule moving in the gas. If the factor were $\frac{1}{2}$ instead of $\frac{1}{3}$, then that quantity would be exactly the kinetic energy. As it stands, the quantity is only $\frac{2}{3}$ of the kinetic energy, and so it must be. If we want to be a bit more precise, we should say that the pressure

$$p = {}^2/_3(\textit{average} \text{ kinetic energy of a particle}) \times N/V ,$$

or
$$p = {}^2/_3(mv^2/2)_{av} \times N/V = {}^2/_3(KE)_{av} \times N/V$$

$$= {}^2/_3(KE)_{av} \times N/V . \tag{21}$$

Historically, the inverse dependence of pressure and volume of a gas was first found by direct experimentation, and is known as Boyle's Law.* It is often written in the form

$$pV = \text{ constant if the temperature is constant },$$

or
$$p_1V_1 = p_2V_2 \text{ at constant T }.$$

This expression implies that there is a relation to be found between the temperature of a gas and the mean kinetic energy of its molecules. We shall find this relation shortly.

Equation **(21)** tells us a great deal about how the gas can do work. It tells us that if the gas pushes against an external pressure only slightly less than its own pressure, then the piston of the working chamber will absorb momentum and energy from the gas molecules, move and increase the volume of the chamber so that the number of molecules per unit volume of gas will decrease. Hence the gas becomes a poorer source of work as the piston expands.

Equation **(21)** also tells us that we have to worry about what happens to the mean kinetic energy of the gas molecule as the piston expands. If we isolate the gas and supply it with no new energy, the average kinetic energy of the gas molecules must necessarily drop,

---

* Called Mariotte's Law on the European continent, and first discovered by Richard Towneley; *c.f.* D.S.L. Cardwell, *From Watt to Clausius* (Cornell University Press, 1971).

---

as the piston expands. In that case, the gas loses working capability by virtue of a falling mv^2 factor, as well as by virtue of a density that drops, as the expansion occurs. Alternatively, we might supply enough energy to keep the average molecular kinetic energy constant as the gas does its work. These are two of the choices we can make in order to deal with the working gas system. We shall have occasion to treat these two cases in some detail.

Although Eq. **(21)** is a relation among the variables p, V and N, which are variables we want to use, this equation still contains a molecular quantity, the average kinetic energy of the particles, which is a microscopic rather than a macroscopic variable. To replace this quantity with the kind of macroscopic variable we want, we must turn to another experimental relation concerning pressure and, this time, *temperature*, in a system of constant volume. The relation is now known as Charles' Law,* and has the form

$$\text{pressure} \quad \propto \quad \text{temperature at constant volume} ,$$

or $\qquad p = (\text{temp} + \text{constant}) \times \text{constant, when V is fixed} .$ **(22)**

The constant inside the parentheses can be incorporated directly into the definition of the temperature scale. Thus, if temperature is measured in degrees centigrade, then the form of Charles' Law is

$$p = [T(^\circ C) + 273] \times \text{const} ,$$

but if we define a new scale, the *Kelvin* (abbreviated "K") or absolute scale, whose zero is at $-273^\circ C$, we can write Charles' Law in a simpler and presumably more significant way. Let

$$T(^\circ K) = T(^\circ C) + 273 ;$$

---

* Discovered by Joseph-Louis Gay-Lussac in 1808; J.A.C. Charles apparently believed it was *not* a general law. See Cardwell, loc. cit.)

---

then $\qquad\qquad p = T(°K) × const$,

when V and N are constant. Explicitly, we may now write the pressure as

$$p = T × N/V × k_B \ , \qquad\qquad (23)$$

where k_B is the constant of proportionality, known as Boltzmann's constant, after Ludwig Boltzmann. This expression includes all the relationships that our generalizations have told us.

Now compare Eqs. (21) and (23). From the former,

$$pV/N = {}^1/_3(mv^2)_{av} = {}^2/_3(KE \text{ per molecule})_{av} \ ; \qquad (24)$$

from the latter, $\qquad pV/N = k_B T$. $\qquad\qquad (25)$

Note that **temperature** was introduced into our discussion with no definition except that it is what we read on a thermometer. Tacitly, we assumed that it was an arbitrary but reliable measure of the intensity of heat, and that it can be quantified by measurement with a capillary of mercury or alcohol whose height varies with temperature. So construed, temperature is clearly a "macro" variable, but its meaning seems more arbitrary than those of pressure or volume, for example.

Now, if we are to make Eqs (21) and (23), or better, Eqs. (24) and (25) consistent, we must *identify the* **absolute temperature** *T as a measure of the average kinetic energy per molecule in the gas.* We must require that

$$k_B T = {}^1/_3(mv^2)_{av} = {}^2/_3(KE/molecule)_{av} \ . \qquad (26)$$

This identification immediately changes temperature into a very precisely defined "macro" variable with a sharp interpretation at the "micro" level. The recognition that temperature measures the mean kinetic energy per molecule is a basic concept: it pervades all parts of science that connect the micro and macro levels.

The value of the constant k_B, is 1.38×10^{-23} joules per °K. Because Boltzmann's constant is very small and N is typically a very large number, it is useful to rescale our numbers to deal with quantities closer to unity. The rescaled quantities we shall now describe were actually used long before Boltzmann's constant was evaluated. However, the scale relation is the quantitative key that connects the microscopic and macroscopic worlds.

The way we rescale is by counting particles in units of many particles at once, instead of individually. Like counting by twos or dozens, we can count particles so many at a time that the numbers N and k are replaced by numbers near unity, usually symbolized by n and $\mathbb{R}$. How do we pick this counting number? The easiest and now-standard way is to use the number of atomic mass units per gram, called Avogadro's number N_A, as the standard counting number. Hydrogen atoms each weighs approximately one mass unit. (Strictly, the isotope ^{12}C of carbon is used to define the scale of atomic mass units: the mass of this isotope is taken as precisely 12.) Hence N_A hydrogen atoms weigh approximately one gram. Therefore, if we count particles N_A at a time, and we deal with macro (gram-like) quantities, then we will have numbers of order unity to deal with in our quantity n. We define the counting unit as the *mole*:

one mole of articles $= N_A$ of these articles .

We obviously require that

$$Nk_B = n\mathbb{R} \ , \tag{27}$$

so that if we count articles N_A at a time and thereby define the number of moles

$$n = N/N_A \ , \tag{28}$$

then we must, correspondingly, define

$$\mathbb{R} = k_B N_A \ . \qquad\qquad (29)$$

The constant $\mathbb{R}$ is called the **gas constant**.

Avogadro's number is approximately 6.02×10^{26} atomic mass units (amu) per kilogram or 6.02×10^{23} amu per gram. Hence the gas constant is approximately 8.3 joules per mole, per °K. We shall soon express $\mathbb{R}$ in other units as well.

Finally then, the equation of state for an ideal monatomic gas, expressed in terms of macroscopic variables, is

$$pV = n\mathbb{R}T \ , \qquad\qquad (30)$$

which contains Boyle's Law, Charles' Law and the dependence of all the macroscopic variables of the state of the gas.

Note that there are two sorts of variables in our expressions. The pressure p and temperature T are measures of intensity, while V and n or N are measures of quantity. The former, such as the force *per unit of area,* or the *average kinetic energy per molecule*, are called *intensive* variables, and the others, that measure amount of material or amount of space, for example, are called *extensive* variables. Intensive variables do not depend on the quantity of matter in the system. Extensive variables depend directly on the amounts.

Problems

1. Estimate the temperature at which the mean speed of an oxygen molecule (mass 32 atomic mass units) is 10^5 cm/sec or 1000 m/sec or 1 km/sec. How many joules of kinetic energy (of translation) does 1 mole of oxygen contain at this temperature?

2. Estimate the kinetic energy of translation, that is, of free motion, in joules, of one molecule of oxygen (32 amu) in air at 300°K (slightly warmer than room temperature). How many joules of kinetic energy are contained in the translational motion of one *mole* of air? (Assume the average molecular weight is 30 amu and that the temperature is 300°K.) What is the average speed of an air molecule at room temperature? Compare this with the speed of sound in air at room temperature and with the speed of these common macroscopic objects: automobiles at the interstate highway speed limit; bullets; airplanes.

3a. Show in terms of dimensions, that the rate of transfer of momentum per unit area, per second, to a wall is equal to the pressure on the wall.

  b. The heliometer is an engine driven by light. It consists of several light-weight vanes, usually four, attached to a central axle. One side of the vanes is black and the other is silver. The assembly of vanes and axle is supported in a sealed, transparent container. If the container is evacuated until it contains an exceedingly small quantity of air, less than 10^{-6} atm, the pressure of light alone causes the vanes to turn the axle. How does this happen? Note that black things absorb light and shiny things reflect light. Which way does the device turn? A sketch of the device is shown; draw your own to illustrate how the heliometer works. If a little air is present, perhaps 10^{-1} or 10^{-2} atm, the heliometer also turns, but in the direction opposite to its direction when the container is exhausted. What drives the heliometer in this partially evacuated condition? (This last is the most subtle part of the problem.)

4a. What volume of air at 1 atm of pressure and 20°C is needed to fill a bicycle tire whose volume is 1 liter, to a pressure of 70 pounds/in^2 (psi), the unit on most American tire pressure gauges, or 5 atm, at 20°C? (1 atmosphere of pressure is equivalent to 14 psi.) Note that the 70 psi is the excess pressure above the one atmosphere in the tire before we begin filling it. We pump up the tire by filling the cylinder of a hand pump with air at 1 atm and then, by compressing the gas in the cylinder, adding all the air in the pump to the air in the tire. If the pump has a volume one third that of the tire, how much is the pressure in the tire above 1 atm after three full strokes of the pump?

b. A tire valve is held shut by the higher pressure of the air inside the tire and only opens when the pressure outside is greater than the pressure inside the tire. When the bicycle tire pressure is 4.5 atm, how far down the 50 cm long cylinder of a hand pump must you push the piston before the valve opens? Explain why long bicycle pumps are advantageous over short bicycle pumps. What is the consequence of the "dead" volume between the valve in the tire and the lowest point that the piston can reach?

5. In interstellar space, the gas is essentially all hydrogen. In many regions, it is in the form of hydrogen <u>atoms</u>. The temperature of this gas is now believed to be 3°K. Its density is approximately 1 <u>atom</u> per cm^3.

   a. What is the pressure of this gas?

   b. What volume of interstellar space contains 1 gm of matter?

6. At the top of Mt Everest, the mean pressure is only 210 torr (760 torr = 1 atm) and the mean density is 0.425 kg/m^3. What is the mean temperature? Compare this temperature with normal room temperature. Also compare the quantity of oxygen available for breathing with that at sea level, 1 atm pressure and 273°K. (How should "quantity" be defined?)

7a. An automobile produces about 2 kg of carbon monoxide, CO, per average day of driving. What is the pressure of this much gas, in atm, in a volume 50 km (50,000 meters) × 1000 meters^2, roughly the volume associated with the roadway traveled by the car in an average day? Assume T = 300°K. (Note: $\mathbb{R}$ ≈ 0.08 liter atm/°C.)

   b. The concentration of CO in the Callahan Tunnel, Boston, was found to be 50 ppm (parts per million*) in one set of measurements. How many cars were required to produce that concentration (or that pressure, 50-millionths of an atm)? Base your answer on 7a.

---

* ppm means molecules per million molecules picked at random.

---

8. A producer is planning a new film to combine nostalgia and disaster, in which he will reenact the burning of a giant balloon filled with pure hydrogen. The balloon is elastic, and virtually weightless. Like all non-rigid craft, the balloon expands as it rises. Why? The maximum altitude the balloon will reach will be 10 km, at which the pressure of the surrounding atmosphere is 0.3 standard atmospheres or 30 newtons/meter^2, and the temperature is 250°K. The balloon can stretch to a volume of 30 million liters (3 × 10^7 l or 3 × 10^4 meters^3). How many moles of hydrogen are required to fill the balloon? What will the volume of hydrogen be when the balloon is filled at the earth's surface, where the pressure is 1 standard atmosphere and the temperature is 300°K? [You will have to anticipate that for an ideal or nearly ideal gas, pressure × volume, pV, are equal to the number of moles (amount of gas, counted by number of molecules, 6 × 10^{23} at a time) × absolute temperature T times a conversion factor $\mathbb{R}$: pV = $n\mathbb{R}$T. $\mathbb{R}$ is approximately 8 joules/mole - degree.]

5. THE EQUATION OF STATE AND THE REPRESENTATION OF STATE CHANGES AND WORK

In Chapter 4 we established the connection between the pressure p, volume V, temperature T and amount of material (n moles or N particles) in a hypothetical gas that is described by these conditions:

a)   the gas is composed of point-masses, each of whose mass is m;

b)   the particles undergo elastic collisions with the walls and exert no forces on each other.

We introduced other assumptions just to simplify the algebra, without changing the physical picture significantly. The resulting equation, written

$$pV = NkT \qquad (25)$$

or
$$pV = n\mathbb{R}T , \qquad (30)$$

is a very good representation of the behavior of gases under most conditions. The more rarefied the gas, the more accurately its behavior is represented by Eqs. (25) and (30). It is particularly significant that these equations make no reference to the nature of the gas or even to the mass of its component particles.

When a gas is compressed and its molecules are, on the average, brought closer to each other, the forces between these molecules can no longer be neglected. Both the attractive forces that make molecules condense into liquids and solids, and the repulsive forces that keep the molecules from collapsing into each other, cause real gases to deviate from the behavior described by Eq. (30). These deviations vary from substance to substance, as we might expect. If a gas is cooled and compressed enough, it condenses to form a solid or a liquid, and clearly the liquid and solid forms of substances differ, one from the other.

Gas is a state of matter in which the nature of the particles is almost irrelevant to the equation of state. Its properties depend only on the size of the particles being very small, compared with the volume in which they move and with the distances between these particles. The equations of state for liquid and solid states cannot, in general, be written in any form valid for all substances with a single universal constant such as k_B or $\mathbb{R}$. The simplest form for the equation of state of a liquid or a solid depends on the compressibility, κ, which is the fractional decrease in volume due to an increase in pressure, and on α, the coefficient of thermal expansion or the fractional change in volume due to an increase in temperature. Thus, if the temperature is unchanged but the pressure varies by an amount Δp, the corresponding fractional change in volume, $\Delta V/V$, is given by

$$- \Delta V/V = \kappa \Delta p$$

(with a negative sign so that κ is a positive number). In terms of the volume V_0 at some standard pressure and temperature, and for $V \approx V_0$, this becomes

$$V = V_0(1 - \kappa \Delta p) .$$

Similarly, if the pressure is fixed but the temperature varies, the fractional change in volume is

$$\Delta V/V = \alpha \Delta T,$$

or, again, in terms of a standard volume of reference V_0, taken at a selected standard temperature and pressure, the volume is

$$V = V_0(1 + \alpha \Delta T) .$$

These two relations together give us a simple but reasonably accurate equation of state for a liquid or a solid. The volume V is connected to the temperature and pressure by the relation

$$V = V_0(1 - \kappa \Delta p + \alpha \Delta T) \ . \qquad \textbf{(31a)}$$

The quantities κ and α are very specific for each substance. They are not constant; they vary with temperature and pressure. However their variations are small and can frequently be neglected. The reference volume V_0 is also a specific number for each substance, and, as we have written our equations, depends linearly on N, the number of molecules in the sample. To remove that dependence, we can replace the constant V_0 with the volume of a standard amount of the substance, e.g. the volume of one mole, which we may call $\mathbb{V}_0$. Then the total standard volume V_0 can be written as $n\mathbb{V}_0$, where n is the number of moles, and Eq. **(31a)** becomes

$$V = n\mathbb{V}_0(1 - \kappa \Delta p + \alpha \Delta T) \ . \qquad \textbf{(31b)}$$

Even though $\mathbb{V}_0$, κ and α depend on the particular substance, in contrast to the one universal constant $\mathbb{R}$ appearing in *all* ideal gas equations, the physical picture represented by Eq. **(31a)** or **(31b)** is quite different from that presented by Eq. **(30)**. In particular, the volume of a gas varies inversely with pressure, while the volume of a solid or a liquid hardly varies at all with pressure; typically, the constants κ and α are only fractions of a percent, per atmosphere or per degree. Thus because solids and liquids are nearly incompressible, they are not generally useful substances for transforming molecular kinetic energy into work; they simply cannot expand enough to achieve large volume changes.

Let us return to gases. The changes that occur when a fixed quantity of gas changes its state can be represented in terms of two properties, such as pressure and volume. If we fix those, we need not, indeed we cannot, specify the temperature arbitrarily because it is fixed by the Equation of State. This allows us to make a very useful graphic representation of the succession of states of the system when it undergoes some process. We do this by recognizing that, for a fixed quantity of gas, the variables p and V (or p and T, or V and T) can be treated like two coordinates in a plane. Specifying p and V is like specifying the x and y coordinates of a point on a graph. For fixed n, each specification of p and V also corresponds to a *state* of the gas, and fixes T for that state. Hence, each point in a p, V plane corresponds to a *different state* of the fixed quantity of gas. Two such states are illustrated in Figure 7, by points I and II.

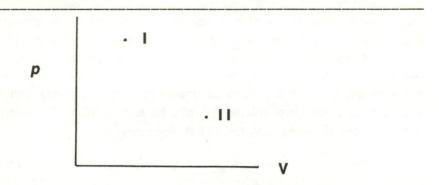

Figure 7. Two states, I and II, for a fixed quantity of gas. State II has the larger volume and lower pressure.

The continuous transformation of a gas from one state through a succession of intermediate states is represented as a curve or line in our p,V graph. For example, in such a graph, a process carried out at constant pressure corresponds to a horizontal line, and a process carried out at constant volume corresponds to a vertical line. These two processes are illustrated in Figure 8.

Note that the processes shown in Figure 8 require that we increase the temperature of the gas, in order to increase the volume in the first process, or to increase the pressure in the second. We shall return to these curves shortly, when we discuss work. A diagram showing the path followed by the system as a curve in a p, V graph is called an indicator diagram. The name derives from the simple manner in which James Watt and his colleagues connected the pistons and pressure gauges of steam engines to draw such diagrams automatically as the engine ran, thus showing the states of the gas in the engine.

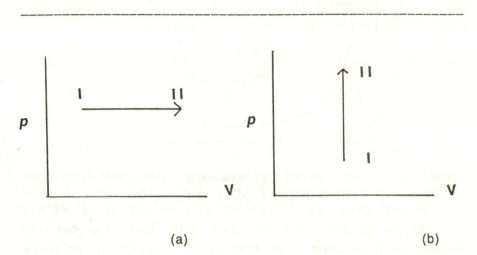

(a)   (b)

Figure 8.   Two types of processes for a gas: a) a process carried out at constant pressure, from a low volume, at I, to a high volume, at II; b) a process carried out at constant volume, from a low pressure, at I, to a high pressure, at II.

There are other ways to change the state of a given quantity of gas, of course. One is to change the volume of the gas while holding its temperature *constant*. Because the product pV is equal to $n\mathbb{R}T$, in this particular process, called an **isothermal** process, the pressure varies inversely with the volume: $p = constant \times 1/V$. This relation describes a hyperbola, as shown in Figure 9.

The increments of work, done by small changes in value, as we saw in Eq. **(17)**, are each given by the product of the pressure and the small increment of volume. If the pressure is constant, as in Figure 8a, then the total work going from I to II is the constant value of the pressure, multiplied by the change in volume. This is just the height of the line, multiplied by the length of the base of a rectangle, or simply the area of the rectangle, as shown in Figure 10.

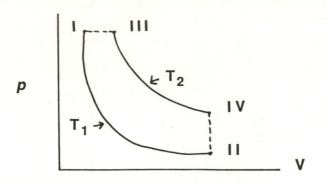

Figure 9. Two curves representing isothermal (constant temperature) transformations of a gas. The curve marked T_1 has the same initial pressure at I as the curve marked T_2, which is presumed to start with state III, but the volume at I is lower than that at III. Hence T_1 must be lower than T_2. The final volumes at II and IV, respectively, are the same, but the pressure for T_1 is lower than that for T_2, so again, we see that T_1 is less than T_2. The curves are portions of hyperbolas.

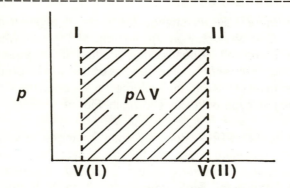

Figure 10. The work done in a constant-pressure or isobaric process is the area p [V(II) – V(I)] or $p\Delta V$.

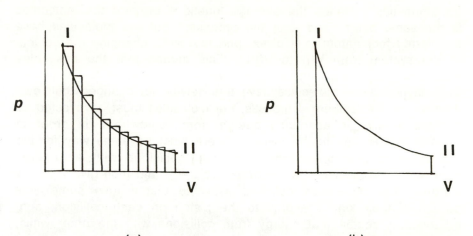

(a)   (b)

Figure 11. Areas representing work carried out in an isothermal process: a) a close approximation, the sum of areas of rectangles, as in Eq (17a); b) the exact representation, the limit of very narrow rectangles in (a).

In a process carried out at constant volume (called an **isochoric** or **isometric** process), there is no mechanical work done, since ΔV is zero. In a process in which the pressure varies with volume, such as an isothermal process, we must carry out a sum over many small steps, as in Eq. **(17a)**. The limit of very narrow rectangles, or very small increments of volume, is indicated in Figure 11b, where the area under the curve is precisely the work carried out when the gas moves from state I to state II. A process carried out with the pressure constant is called an **isobaric** process.

If work is carried out according to an isobaric or an isothermal process, what actually occurs at the molecular level, and what must we do to obtain this work? In the isothermal process, the system has the same temperature at the beginning and end of the transformation. Hence the average kinetic energy of the molecules is the same before and after the process. But the molecules have transferred momentum to a piston and moved it, changing the volume of the system from V(I) to V(II). This means that the molecules must have put energy into whatever we have connected to the piston, even though these molecules have themselves not changed their own energy. In the isothermal process, the molecules must have acted as carriers of energy, absorbing energy from somewhere, in order to transfer that energy to the piston. There must be some way for the molecules to pick up energy from their collisions with an energy supply – – which might be the *other* walls – – and then give it to the piston. The molecules colliding with the piston lose some little portion of their kinetic energy to the piston on each collision, and, on average, regain that energy from collisions with the other walls, which remain at their original fixed temperature. The energy source that keeps those walls at a constant temperature must be the source of the energy carried by the molecules to the piston. We call this source a *thermostat*, because it holds the temperature constant. (This should not be confused with the thermal regulator that often goes by the same name.) The thermostat is the energy source for isothermal work.

In the case of the isobaric or constant pressure process, the gas ends its process with the same pressure but a larger volume than it had originally. In this case the gas not only transports energy, equal to the area under the p-V curve, and turns this energy into work; it also must increase in temperature during the process in order that the product pV increases. Hence we must supply not only the energy equal to the work $p\Delta V$, we must also add enough energy to increase the temperature of the gas from $T(I) = pV(I)/n\mathbb{R}$ to $T(II) = pV(II)/n\mathbb{R}$. A thermostat does not suffice to carry out an isobaric process. We need a heater, to *raise* the temperature.

In an isochoric or constant volume process, where no work is done, the only change in the gas can be a change in the average kinetic energy of the molecules. In a change from state I to state II, as in Figure 8b, we merely raise the temperature, by raising the molecular energy.

One other important class of change deserves our attention. This is the process in which work is extracted from a gas into which we supply no extra energy. Such a process is called *adiabatic*. If we extract work by letting a gas push a piston, and do not keep replacing that energy, then the work must come from the molecules themselves, each giving up a tiny fraction of its kinetic energy with every collision at the piston. Slowly the average kinetic energy of the gas molecules must fall, as must the ability of the molecules to do more work. If an isothermal process and an adiabatic process begin at the same state the final state of the adiabatic process must fall on a point below the final point of the isothermal process in a *p*-V diagram. This is illustrated in Figure 12. The result is that the adiabatic process does less work than the isothermal, if they start at the same state and end at the same volumes.

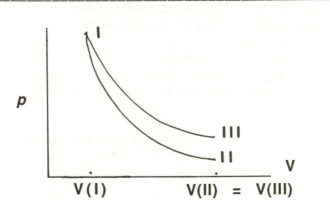

Figure 12. Curve I-III represents an isothermal process, and curve I-II, an adiabatic process starting from the same state. Note that point II corresponds to a lower temperature than point III, and the area under I-III is greater than that under I-II, even though the final volumes are identical.

Problems

1. All engines that convert heat into work operate on some kind of *cycle*. That is, they begin in a state corresponding to one point, call it I, in the p,V plane and pass through a succession of other stages, tracing out a path in the p,V plane, and then return to point I. One example is the cycle corresponding to the operation of the conventional internal combustion engine in an automobile with its four-cycle (really four-step cycle) engine; this cycle is called the Otto cycle. Another, and perhaps the simplest conceptually, is the Carnot cycle, conceived and systematized by the French engineer Sadi Carnot. The Carnot cycle consists of an isothermal expansion from I to a second point II, an adiabatic expansion from II to a third point III, an isothermal compression from III to a point IV, and finally an adiabatic compression IV back to I. Sketch the Carnot cycle, using arrows to show the direction for each step; indicate (e.g. by shading) the work done *by* the system in the first two steps, the work done *on* the system in the third and fourth steps, and the net work done by the system in the cycle.

2. An engine that uses an ideal gas as working fluid is made to operate on the tri-cycle, a three step cycle. The engine starts at state 1, with a pressure p_1, a volume V_1 and a temperature T_1 of 500°K. In the first step, the gas expands until its volume is doubled to $2V_1$, and its pressure is halved to $^1/_2 p_1$, along a straight line to state 2 in the indicator diagram. [Incidentally, the equation for the line is $(p/p_1) = -\,^1/_2 (V/V_1) + \,^3/_2$.] The second step is a contraction at constant pressure from state 2, where the volume is $2V_1$, to state 3, where the volume is again V_1 but the pressure is still $^1/_2 p_1$. In the third step, the gas is heated at constant volume until its pressure returns from $^1/_2 p_1$ back to p_1, so that the gas goes from state 3 to state 1, completing a cycle.

a. Draw an indicator or p-V diagram for this cycle, being careful to show the quantitative relations among the pressure and volumes at the three vertices.

b. What is the temperature of the gas at the <u>end</u> of the <u>second</u> step?

c. Show on the indicator diagram how much work is done in one cycle.

d. Suppose that the tri-cycle engine just described contains one mole of gas as its working fluids. Show that the work done in one cycle by this engine is equal to $\mathbb{R}T_1/4$ or 1000 joules.

3. The Otto cycle, describing the behavior of the conventional internal combustion engine of the automobile, can be represented approximately by the following four steps:

   i) a reversible adiabatic compression;

   ii) heating at constant volume (representing the combustion step);

   iii) a reversible adiabatic expansion, during which the hot gas does work on the driveshaft, and

   iv) cooling at constant volume to return to the original state.

a. Sketch the Otto cycle in a p-V indicator diagram and indicate the net work carried out in one cycle of the engine by suitably marking the graph.

b. The maximum volume in a particular 1-cylinder Otto cycle engine is 80 cm^3 and the minimum volume is 20 cm^3. The maximum temperature is 1800°K. The minimum temperature is 400°K, and is reached when the pressure is 1 atm. How many moles of gas does the cylinder hold?

c. What is the pressure in this engine, in atmospheres, when the temperature is 1200°K, at the end of the adiabatic expansion step?

6. HEAT CAPACITIES OF GASES

How much energy is required to cause a specified change of state – – for example to achieve a given volume change in a constant pressure or constant temperature process? To answer such a question for gases, the simplest systems, and then for solids and liquids, we must examine the *molecule* again, and introduce a general, quantitative and powerful property known as *heat capacity*. The heat capacity of anything may be defined as the amount of energy that an external source must supply in order to increase its temperature one degree, by a specified kind of process. The heat capacity, like work, is indeed process-specific. Historically, it was first defined as the amount of *heat* that must be absorbed by something to increase its temperature one degree, and is still often defined this way, but we have not yet examined the meaning of heat, so we use the more general definition for the present.

The first process we consider is the heating of a fixed amount of gas in a fixed volume. We suppose, from this point on, that k_B and $\mathbb{R}$ are in units of joules/degree Kelvin and joules/mole-degree Kelvin, respectively. The temperature of the gas is related to the average kinetic energy of the molecules by the condition that

$$k_B T = {}^2/_3 (KE/molecule)_{av}$$

or, if we count by moles instead of molecules,

$$N_A k_B T = n\mathbb{R}T = {}^2/_3 (KE/mole)_{av} \ .$$

Rearranging, we have

$$(KE/molecule)_{av} = {}^3/_2 k_B T$$

or

$$(KE/mole)_{av} = {}^3/_2 \mathbb{R}T \ .$$

If the only kind of energy the gas molecules have is their kinetic energy, then they require an addition of $^3/_2 k_B \times 1$ joules/molecule or $^3/_2 \mathbb{R} \times 1$ joules/mole to exhibit a rise of one degree in temperature. Hence the *heat capacity* for a process carried out at constant volume is $^3/_2 k_B$ joules per degree, per molecule or $^3/_2 \mathbb{R}$ joules per degree, per mole, for these simplest of gas particles. We denote the heat capacity of one mole of a substance, for a constant volume process, by C_V.

For a solid or a liquid, it makes little difference whether one measures its heat capacity at constant volume or in some other way, such as at constant pressure which lets the sample expand and do work. For a gas, by contrast, the energy absorbed in a process carried out at constant pressure is considerably greater than that absorbed in a constant volume process because of the work the gas does in expanding. In addition to requiring an amount of energy $^3/_2 \mathbb{R}$ for each degree of rise of temperature, (for each mole of gas), the gas also extracts $p\Delta V$ from its source of energy and puts it into work done on the external source of pressure, whatever it is. If the gas expands by ΔV at constant pressure, this work is $p\Delta V$. But the work of expansion $p\Delta V = n\mathbb{R}\Delta T$, so the n moles of gas must absorb an extra $n\mathbb{R} \times 1$ units of energy (joules or ergs) for every degree of increasing temperature associated with the work the gas does. Hence the total amount of energy the source must supply to each mole to increase the temperature one degree, in an isobaric process, is $^3/_2 \mathbb{R} + \mathbb{R}$. The first term, $^3/_2 \mathbb{R}$, represents the true heating, and the second, $\mathbb{R}$, the work done. The energy represented by first term stays in the gas; that in the second is transferred to the piston or other external source of pressure. The heat capacity of one mole of our ideal monatomic gas at constant *pressure* is $^5/_2 \mathbb{R}$, usually called C_p.

Until now we have neglected any internal structure in the molecules. If they are in any way capable of storing energy internally, for example in vibration or rotation, then our simple model of heat capacity is inadequate. The model still describes the translational kinetic energy accurately, but we must allow energy to flow into the other kinds of motion of the molecules. If a particle has internal structure, then every internal motion – – rotation or vibration – – may hold energy. For example, a molecule made of two atoms bound together – – a *diatomic* molecule such as nitrogen, N_2, or oxygen, O_2 – – may vibrate, or it may rotate about either of two (arbitrary) axes that are perpendicular to each other and to the line joining the centers of the two atoms.

If a system is in equilibrium, then the average energy associated with each type of motion of the particles is constant in time. Let us anticipate some of our later discussion and recognize that, at equilibrium, the mean kinetic energy associated with all types of motion must be the same. This statement means that the effective temperature associated with all types of motion – – translational, vibrational or rotational, for example – – must be the same in a system at equilibrium. Our notion of thermal equilibrium dictates such a condition: a system is only in equilibrium if all its parts are at a common temperature.

In the ideal monatomic gas, each molecule has on the average $^3/_2 k_B T$ of kinetic energy, which is distributed into motion along the three independent axes, x, y, and z. Hence on the average each independent degree of freedom of translational motion, i.e. motion in each of these three independent directions, has $^1/_2 k_B T$ of kinetic energy. Similarly, the average kinetic energy each degree of rotational freedom in a more complex molecule must carry is $^1/_2 k_B T$. A diatomic molecule can rotate about either of the two independent axes perpendicular to the line of the two atoms, so it has two degrees of freedom for rotation.

Likewise each degree of vibrational motion carries an average of $^1/_2 k_B T$ of kinetic energy. Vibrational motion, in contrast to rotational and translational motion, also has potential energy. (Remember our discussion of ideal springs.) The vibrator continually exchanges kinetic and potential energy. In the perfect harmonic oscillator, the average amount of energy in the form of potential energy is exactly equal to the average amount in the form of kinetic energy. Consequently a vibrating species has $^1/_2 k_B T$ average energy in the form of kinetic energy and $\frac{1}{2}k_B T$ in the form of potential energy, when it is in equilibrium.

This argument tells us that a diatomic molecule holds $\frac{3}{2}k_B T$ in translational motion, $2 \times \frac{1}{2}k_B T$ in its two degrees of rotational motion and $\frac{1}{2}k_B T + \frac{1}{2}k_B T$ the kinetic and potential parts of its one degree of vibrational motion, giving a total energy of $\frac{7}{2}k_B T$ in each molecule. If we wish to increase the temperature of a gas composed of diatomic molecules by one degree, then we must give each molecule ($\frac{7}{2}k_B$ joules/degree) $\times$ 1° of energy (taking k_B in units of joules/degree), or give each mole $\frac{7}{2}\mathbb{R}$ joules (with $\mathbb{R}$ in joules/mole-degree), provided we maintain the volume constant so that the gas does no work. Thus the constant volume heat capacity C_V of a diatomic gas is $\frac{7}{2}\mathbb{R}$ joules/mole-degree. The constant pressure heat capacity C_p is greater than this by the amount of work done by one mole of gas in increasing $p\Delta V$ in the amount corresponding to a one degree rise in temperature, or (1 mole) $\times$ $\mathbb{R}$ joules/mole-degree 1° = $\mathbb{R}$ joules This means that the heat capacity $C_p = C_V + \mathbb{R}$ or $\frac{9}{2}\mathbb{R}$. The difference between C_p and C_V for a gas is, as we can now see, always $\mathbb{R}$, regardless of the complexity of the gas.

To complete our survey of the heat capacities of gases, we take only one more step, and estimate the heat capacity of a gas of arbitrarily complex molecules. Suppose a molecule contains n atoms; how many translations, rotations and vibrations does this molecule have? Clearly, every molecule has exactly three translations, corresponding to the motion of its center of mass along the three independent directions, x, y and z. If the molecule is nonlinear, it can rotate about any of three independent axes, and therefore has three rotational degrees of motion. If the molecule is linear, then, like the diatomic molecule, it can only rotate about two axes, since essentially all the mass falls <u>on</u> the bond axis of the molecule, the line containing the atomic nuclei. All the other degrees of freedom are vibrational.

How many other degrees of freedom are there? To answer this, we ask how many degrees of freedom there are altogether. If there are n atoms, and we make our molecule by bringing these atoms together from great distances, then each atom initially has 3 independent degrees of freedom, giving us 3n in all. These degrees of freedom are by no means lost when the molecule forms. They are only transformed, in part, from translational to vibrational and rotational degrees of freedom. In the stable molecule, 3 must remain translational in the stable molecule, 2 or 3 (depending on whether the molecule is linear or nonlinear) must be rotational, and the other 3n − 5 or 3n − 6 degrees must all be vibrational, since that is the only possibility left.

Thus, the constant volume heat capacity, C_V, per mole, due to atomic motion in the molecules of gas, each containing n atoms, must be

$$C_V = \tfrac{3}{2}\mathbb{R} + \tfrac{3}{2}\mathbb{R} \begin{array}{l} \mathbb{R} \text{ (linear)} \\ \text{(nonlinear)} \end{array} + \begin{array}{l} (3n-5)\,\mathbb{R} \text{ (linear)} \\ (3n-6)\,\mathbb{R} \text{ (nonlinear)} \end{array}$$

$$\text{translation} \qquad \text{rotation} \qquad\qquad \text{vibration}$$

The constant pressure heat capacity C_p is larger by the amount $\mathbb{R}$, per mole, as in the case of the monatomic or diatomic gas.

Atoms themselves have internal structure, with light, negatively charged electrons normally in stable dynamic states attached to heavy, positively charged nuclei. Electrons are capable of being excited and even of being set free, provided they are given sufficient energy. In hot environments such as the sun or electric arcs, these energies are available; however, at temperatures we ordinarily experience, the available energies are insufficient to unlock the electrons from their tightly-bound, stable states, so the electrons do not contribute measurably to heat capacities of free molecules.

Problems

1. Molecules heated to sufficiently high temperatures break into their constituent atoms. Consider a collection of identical, nonlinear molecules, each containing n atoms. What change occurs in the heat capacity, per molecule of the system, as a result of breaking the molecules completely into free atoms?

2. Electrons in atoms and molecules are capable of absorbing energy and thereby of being put into excited states. The release of this excitation energy is the light given off in a spectral line when an electron returns to its lowest available state. Despite this capability, the electrons were neglected in our development of heat capacities and our results fell into good agreement with real observations. Explain why the electrons do not contribute significantly to heat capacities of atoms and molecules under ordinary conditions. Under what conditions, other then in stars and electric arcs, might their contribution be more important?

3. The water molecule, H_2O, is triangular. Carbon dioxide, CO_2, is linear. How many calories are required to heat one mole of water vapor from 300°K to 1000°K? How many calories are required to heat one mole of carbon dioxide from 300°K to 1000°K? (Assume the heating is done at constant pressure.)

7. HEAT CAPACITIES OF SOLIDS: THE NATURE OF HEAT

The heat capacities of solids and liquids can hardly be expected to contain a large component arising from translational motion and comparable amounts from rotation and vibration. In a solid, molecules simply are not free to move translationally at all. In fact, a best first guess about the heat capacity of a solid is based on the simple notion that it is just a giant molecule. This assumption requires that the entire solid have only $^3/_2 k_B T$ of translational energy – – not $^3/_2 \mathbb{R} T$ – – because only the center of mass of the entire solid is free to move translationally. If the solid sample consists of N atoms, then it has 3N degrees of freedom in all, and 3N – 3 are locked into the motions of atoms around specific lattice positions in the solid. Since N is of order 10^{19} or more, the 3 subtracted degrees of freedom make an infinitesimal, immeasurable change in the total number 3N, and we can say that there are 3N degrees of *vibrational* motion for the N atoms. Each atom then has $3kT$ total energy, of which $^3/_2 k_B T$ is kinetic, and $^3/_2 k_B T$ is potential, on the average. Hence the total vibrational energy is, according to our model, $3k_B T \times N$ or $3n\mathbb{R} T$.

The value of $\mathbb{R}$ is $N_A k_B$, or $6.02 \times 10^{23} \times 1.38 \times 10^{-16}$ or 8.31×10^7 ergs/mole deg, or 8.31 joules/mole deg (since 1 joule = 10^7 ergs). Thus, our model says that the heat capacity, per mole, of a solid should be 3 × 8.31 joules/degree, or approximately 25 joules/ degree. We can design experiments in which we do mechanical work on a solid and measure both its temperature change and the amount of mechanical energy that produces this change. For example, we could hit our sample with a hammer, or subject it to some other form of mechanical stress. However, it is far easier to carry out the measurement of heat capacity if we recognize a general principle and put the energy into the solid by a different means, specifically in the form of **heat**.

Heat energy is just as much a form of energy as mechanical or any other sort of energy. Heat can be used just as effectively as (and often more easily than) mechanical, electrical or light energy to increase temperature. We can put a measured amount of heat into a solid and measure the change in its temperature that the specific amount of added heat produces. This approach becomes useful just as soon as we know what heat is.

Heat is that kind of energy whose concentration or intensity is measured by temperature. The total amount of heat energy in a sample is proportional both to the value of the intensive variable of temperature, and to the amount of material present. But temperature, we have decided, is a measure of only the translational energy. Surely the total energy that we call heat must include not only the kinetic energy stored in rotations and vibrations but also whatever potential energy is, on average, held in vibrations and in any other pockets the particles may have for storing energy. In other terms, molecules with a large **heat capacity** must store more energy, per degree of temperature, than molecules with a small heat capacity. This is after all why the property is called "heat capacity". How, then, can we relate the heat content of a sample to the energy stored in its various degrees of freedom?

We have been relating the amount of energy in translation to the amounts in vibration and rotation, but merely by quoting the results, without giving a general basis for the claim. Now it is time to recognize explicitly the general relation that justifies our claim. This relation is called the **Principle of Equipartition of Energy** which states that, in a system at equilibrium, all forms of kinetic and potential energy contain the same amount of energy, on the average. To illustrate this principle, suppose we work with a bottle of gas molecules that are capable of vibrating as well as of translating. Suppose we try to put energy only into the translational motion of the molecules. The molecules collide with one another; as a result of those collisions, part of that energy will flow from translational motion into vibrational motion until the average energy in each of the modes of translation is the same as the average kinetic energy of each mode of vibration. But this also becomes the same as the average potential energy of vibration in each mode, and the average kinetic energy of rotation about any single axis. The different modes of storing energy share their energy until they all have the same average kinetic energy. We say, when this occurs, that the energy is *equipartioned*.

In the case of a solid, the Principle of Equipartition assures that the average energy is the same in all the vibrational modes of the entire solid. The Principle of Equipartition says that all the degrees of freedom of a system must share a common temperature when the system is in equilibrium. Incidentally, this Principle tells us nothing whatsoever about how long a system may require to reach that equilibrium.

If heat is the extensive variable that is the counterpart of the intensive quantity we call temperature, if temperature measures the average translational energy or kinetic energy of the particles, and if the Principle of Equipartition is valid, then heat is the measure of the total energy associated with the internal molecular motion of the system, when this internal motion is, on average, in equilibrium and displays a well-defined temperature. This concept of heat assures that any kinetic energy of the system moving as a whole, is excluded from classification as heat.

For historical reason, heat was, for many years, measured conventionally in units of calories, rather than joules. The calorie is defined as the heat capacity, *per gram*, of water at 15°C. This quantity is clearly one that developed from its convenience, that is, from its ease of measurement. It has no natural relation to the usual metric units of energy (such as the joule), any more than the inch has to the centimeter. But if heat is energy, we need to know the scale factor that converts a measurement in one set of units into the other.

Several units besides the calorie have been used to measure quantities of heat and other forms of energy. Often in engineering practice, the "English unit" of heat, the British Thermal Unit or btu, is still used. This is the amount of work required to heat one pound of water precisely 1°F (not 1°C), from 39.1°F to 40.1°F. One btu is equivalent to approximately 250 calories. Essentially, a calorie heats a gram of water (1 cm^3) 1°C, and a btu heats a pound of water (very close to 1 pint) 1°F. Both these quantities, based as they are on the special properties of water, a rather special substance, have definitions that lack the generality of the mechanical units of energy, the joule and the erg.

The unit of heat must now be connected with the mechanical unit of energy. If we had a complete theory of the behavior of water, we could establish this connection on theoretical grounds alone. We have no such theory and therefore must derive our relation from observation.

That heat is a form of energy and not a fluid, and that it has a mechanical equivalent, a constant conversion factor equal to the number of calories per joule, was inferred from heat generated in the boring of cannons by Count Rumford in about 1800; the concept was refined by R.J. Mayer in 1842. The quantitative relation was first established by James Joule,[*] who used falling weights to turn a paddle wheel in a liquid. Joule measured the increase in temperature of a known volume of water or oil due to a known weight falling through a known distance. The equivalence relation is found from experiment to be this: 4.18 joules are equivalent to 1 calorie. Hence the gas constant, $\mathbb{R}$, may be expressed in *calories* per mole degree, thus:

$$\mathbb{R}\,(\text{cal/mole deg}) \;=\; \mathbb{R}\,(\text{joules/mole deg}) \times (4.18)^{-1}$$

or

$$\mathbb{R} \;\approx\; 2 \text{ calories/mole deg} \;.$$

Now we can return to the properties of solids and their heat capacities. The heat capacity of a solid should, from our previous arguments, be 3 × 2 or 6 calories per mole-degree, if our model of the solid as a giant molecule of vibrating atoms is valid. In fact, the heat capacities of many solids, particularly metals, at temperatures

---

[*] See M.H. Shamos, *Great Experiments in Physics*, Henry Holt, New York, 1959; and S.G. Brush, *The Kind of Energy We Call Heat.*, North-Holland, Amsterdam, 1976.

---

near room temperature, are very near 6 cal/mole deg. Some 58 (or more) elements have heat capacities between 5.38 and 6.93 cal/mole deg at 300°K. This fact was recognized well before it was interpreted; the relation is called the **Law of Dulong and Petit**. (Strictly, the empirical law says that the heat capacity is approximately 6.3 cal/mole deg. The reason for the discrepancy is that electrons in metals contribute a little bit to the heat capacity.) The Law of Dulong and Petit is so general that it can be used to estimate the weight of one mole of many kinds of substances, from which one can obtain directly the atomic weights of those substances.

Some materials have heat capacities that fall far short of 6 cal/mole deg at room temperature. Diamond has a heat capacity of only about 1.6 cal/mole deg at 300°K, and the value for silicon at the same temperature is approximately 4.8. However if these materials are heated, their heat capacities increase, apparently approaching the value 6. Substances having anomalously low heat capacities are nearly always substances with high melting points. Evidently some of the heat-containing capacity available at higher temperatures is not used by these materials at 300°K or below. This observation turned out to play a key role in the development of the quantum theory of matter. We shall return to this topic as we develop the ideas underlying quantum theory.

Problems

1. The heat capacity at constant volume C_V and the heat capacity at constant pressure C_p differ considerably for a gas – – by the amount $\mathbb{R}$, in fact. However C_p and C_V are almost the same for solids and liquids. Why? What is the relation between the $(C_p - C_V)$ differences of gases and solids and their Equation of State?

2. Explain how one can determine the atomic weight of a metallic element if one has a sample of the substance, a thermometer, a balance, a source of heat (such as a beaker of boiling water), a graduated cylinder or other volume measuring device, a well-insulated jar such as a "thermos" bottle or a thick styrofoam cup (preferably with a lid) and a supply of water. Assume that the Law of Dulong and Petit is valid.

3. When 1 kg of fuel oil burns, it furnishes 10,500 kilocalories of heat due to the chemical transformation of gasoline and atmospheric oxygen into carbon dioxide and water. That is, the heat of combustion of fuel oil is 10,500 kcal/kg.

a. How many joules/kg is this? Use your conversion table for energy units. The density of fuel oil is very nearly 0.75 g/cm^3, and the heat capacity of water is 1 cal/gram-°C.

b. How much water can be heated by one liter of fuel oil from a typical input temperature of 15°C to a typical temperature of 65°C (150°F) for domestic hot water, if we assume that no heat energy is lost? How much water can be heated by one liter of fuel oil if the hot water heater only used 50% of the fuel oil's energy to heat water?

c. Suppose the water in a 100-gallon (or 380-liter) hot water heater in operation, loses 0.17 kilowatts of heat energy continuously to its surroundings. How much fuel oil, supplying

50% of its combustion energy to heat the water, must be burned, per hour, to replace this lost heat? (Recall that 1 kilowatt = 1000 watts = 1000 joules/second.) This is a rather serious problem. Domestic hot water heating accounts for about 4% of the national energy budget. It is possible to reduce the heat losses from hot water tanks by a factor of 2, simply through the use of one additional layer of insulation. About 10% of the total energy required for domestic hot water goes to keeping the water hot, so this would be reduced to 5% with the better insulation. If this extra insulation were applied everywhere, the national energy budget would drop about 0.2%, a considerable amount.

4. How much energy is supplied to a square kilometer of land by the kinetic energy of the falling raindrops in a heavy rain? Assume that the rain clouds are at an altitude of 1 kilometer (about 3000 ft), and that the accumulated rain has a depth on one centimeter. Neglect any frictional effect of the atmosphere on the raindrops. Estimate the mass of water contained in the rain, and then find its potential energy when it was still in the cloud. This potential energy is all converted to kinetic and then to heat energy. This much was Problem 2 of Chapter 3. Assume that: a) all the energy in the falling rain goes into heating roughly the top two cm of the earth's surface, wherever the rain falls; b) the heat capacity of the earth is about 6 cal/mole deg, or 24 joules/mole deg; and c) the density of atoms is about 1/10 mole of atoms per cm^3. What temperature increase would the earth's surface show as the result of 1 cm of rain, from a cloud 1 km high? (Assume also that the rain merely percolates into the soil after it falls.) Evaporation of water requires 540 cal/g or 2160 joules/g. If the rain were to evaporate, rather than to percolate into the soil, and if it were to draw its energy for evaporation from its own energy of falling and from the earth's surface, would the evaporation lead to a net surface cooling, or would the heat of falling be enough to supply this energy?

5. A sample of an unidentified metallic element weighs 100 gm
   and has a total heat capacity of 12 calories/°C, or 0.12
   cal/gm-°C. What is the atomic weight of the element, to
   within 20%?

8. OTHER FORMS OF ENERGY

There are other forms of energy besides heat and mechanical energy that concern us. These forms include light, electrostatic and electromagnetic energy, and chemical energy. (Light is only a special case of electromagnetic energy.)

Electrostatic energy is probably the easiest to understand in mechanical terms. The force between charges q_1 and q_2, on particles 1 and 2, depends in the same way as the gravitational force on the distance R between the charges:

$$F_{electrostatic} = q_1 q_2 / R^2 . \qquad (34)$$

This is illustrated in Figure 13a. Since q_1 and q_2 may have opposite signs, $F_{electrostatic}$ may be negative – – i.e., attractive. The corresponding electrostatic energy is really the potential energy of charges separated by the distance R. If we take zero on our scale of energy to correspond to the two charges infinitely far apart and without kinetic energy, then the electrostatic energy of the system of charges q_1 and q_2 is

$$PE_{electrostatic} = q_1 q_2 / R , \qquad (35)$$

as shown in Figure 13b. Hence, if q_1 and q_2 have opposite signs, the potential energy decreases as the two charged particles approach. If q_1 and q_2 have the same sign, the potential increases as R decreases, and we must do work to "climb the potential hill" and bring the charges together. There is no known gravitational analogue for the repulsive electrostatic force between like charges.

A second form of electric energy is that of a constant electric current moving in a conductor. This energy is essentially the kinetic energy of the moving electrons that have been accelerated by some change in **voltage** or potential, i.e. in *potential energy per unit of charge*. The electrons in a direct current can be thought of as sliding

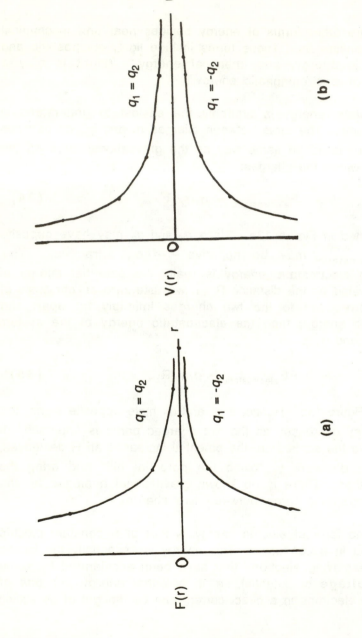

Figure 13. Electrostatic force (a) and potential energy (b) for interacting charges q_1 and q_2, with $|q_1| = |q_2|$.

down a ramp representing their potential energy; the slope of the ramp represents the force on the electrons. The common units of potential energy in such a system are volts × charge; the **volt**, or electric potential, is defined to be the potential energy acting on a unit of charge. This definition is a matter of convenience because of the way we separate the sources of potential, such as batteries, from the charges on which they act. The number of charges (or electrons) passing a given point per unit time is the **current**, whose conventional unit is the **ampere**, with dimensions of charge/time. Hence the quantity

voltage × charge is equivalent to energy

and

voltage × current is equivalent to energy × time^{-1}

or to the amount of energy being delivered, per unit of time. Recall from Chapter 2 that the rate at which energy is supplied is called the *power*, which is measured in watts, kilowatts or megawatts in the metric system or horsepower in English units. One watt is equivalent to the delivery of one joule per second, or 10^7 ergs per second. A common light bulb draws about 100 watts of power, or 100 joules of energy each second, and a large power station generates several hundred million watts of power. The product of the common electrical units of volts × amperes is equivalent to power in watts. Hence a 115 volt line supplying 100 watts of energy to a light bulb is delivering a current of 100 watts/115 volts or about 0.85 amperes.

Voltage and amperage are related by *Ohm's Law* through the property we call **resistance**. This law states that potential in volts = current in amperes × resistance (in units called **Ohms**). Resistance is a kind of electrical friction. The longer a wire is, the greater is its resistance. But the thicker the wire is, the lower is its resistance. It is often useful to characterize wires by their **resistivity**. A wire 50 cm long, and 0.02 cm^2 in area, with a resistivity of 2×10^{-6} Ohm cm has a resistance of $2 \times 10^{-6} \times 50/0.02 = 0.005$ Ohm, which is a typical value for a good conductor such as copper or aluminum.

Electrostatic energy and electric currents are not the only way that energy is associated with electrical forces. The field of force itself, without any charged particles, is a form in which energy can be stored. The electromagnetic field includes the radiation we know as radio waves, light waves and X-rays. Waves of any sort are characterized by the intervals of space and time between points on the wave. These intervals are called, respectively, the **wavelength** λ, the distance between crests at any instant τ, and **frequency** ν of the wave, the time interval between crests at any fixed point in space. The wavelength and frequency of electromagnetic radiation are related through the speed c of the wave fronts:

$$\nu\lambda = c . \tag{36}$$

In empty space, radio waves have lengths of centimeters or meters and frequencies of thousands or millions of oscillations per second. Light waves are much shorter; they have lengths of the order of 5×10^{-5} cm and frequencies of the order of 6×10^{14} per sec; X-rays are shorter and faster still; they have lengths of the order of 10^{-8} cm and frequencies of the order of 3×10^{18} per sec.

An electromagnetic wave makes itself known by its force acting on a charged particle. The electron, in practice, is our most common detector and generator of electromagnetic waves. A radio or television receiving antenna is a device in which the electrons in the metal of the antenna's arms are free to respond to the force of an electromagnetic wave passing through space. A transmitting antenna is one in which electrons are driven back and forth by electric fields applied from the transmitting station. The process of accelerating any charged particle *generates* an electromagnetic impulse; a test charge located some distance away responds to the changing field of force due to the movement of the electron in the antenna. If the antenna's electrons are driven in an oscillatory way, then a test electron in space experiences an oscillatory force field. This is illustrated in Figure 14. The force field of an electron travels from its source, i.e. propagates at a finite velocity – – constant finite velocity c, the universal speed of light – – and not at infinite speed. Consequently, a more distant test charge responds

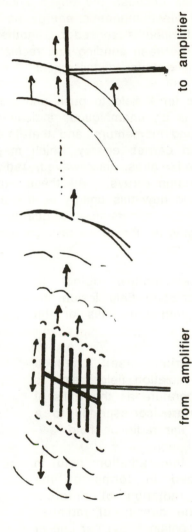

from amplifier

to amplifier

Transmitting antenna with electrons moving back and forth at the frequency of their driving force, determined by the transmitter.

Propagating electromagnetic field in space, with form determined by oscillations of the transmitting antenna.

Receiving antenna with electrons moving back and forth, driven by the force of the electromagnetic field, sending a weak oscillatory impulse to the amplifier of the receiving system.

Figure 14. Transmission of electromagnetic energy through space, as radio waves.

later than a nearer test charge to a given change in the antenna. An oscillatory driving force in the transmitting antenna gives rise to a traveling oscillatory wave moving outward in space. Ordinary radar operates on this basis. The antenna of a radar transmitter sends out pulses of radio waves. Radar operates because the object under observation reflects a little bit of the electromagnetic energy sent out by the radar transmitter. This reflection is sensed by another antenna, the detector. The time delay between sending and receipt, and the known speed of light, tell us the distance of the object.

The ability of the electromagnetic force field to push and pull electrons and other charged particles is an unambiguous indication that this force field transmits energy and momentum, and therefore must store energy. The radiation field carries energy which may have the form of light waves, infrared waves, microwaves, radio waves, ultraviolet waves, X-rays or gamma rays. All these are electromagnetic radiation. To understand how this energy is stored, we shall have to wait until we make an excursion into quantum theory. For the present, we need only recognize certain general principles.

1. In any isolated and insulated hollow object at equilibrium, there must be a radiation field inside the object which is in equilibrium with the walls of the object.

2. The object, being at equilibrium itself, has a well-defined temperature T; the radiation field inside is characterized by the *same* temperature as the walls of its container, and this temperature corresponds to a specific distribution of energy in the radiation field.

3. The distribution of energy in the radiation field in thermal equilibrium is described in terms of the distribution of energy among radiation of different frequencies. One speaks of the density of radiation energy per unit of volume of the container and per unit of frequency range.

Determining the distribution of energy in radiation in equilibrium with (or coming out of) a source at a given temperature was a classic and fundamental problem of physical science in the late nineteenth century. Finding this distribution meant unifying the two monumental fields of physics of the time, the thermodynamics of Ludwig Boltzmann and J. Willard Gibbs and the electromagnetic theory of James Clerk Maxwell.

Lord Rayleigh and James Jeans made one great advance toward joining these two areas by constructing a model for the internal structure of the electromagnetic field. Rayleigh and Jeans supposed that inside an insulated container the radiation field is borne by hypothesized oscillators, like imaginary antennas; the only constraint in the field, they argued, is that the force field of the radiation must be zero at the boundaries of the box. This effectively required all the waves to be like waves in a rope or drumhead, with the ends of the rope or the outside rim of the drumhead fixed and unable to move. Hence the only waves permitted in the container are waves that "fit" the box in the sense of having no field of force to act on a charge at the walls. Figure 15 shows allowed and forbidden waves.

The boundary constraint weighs heaviest on the long waves; for short waves, only slight changes in the wave form (e.g. in percent change in the number of wavelengths in the box) are adequate to bring new allowed waves into the box.

Now we compare the real distribution of energy among frequencies, and the distribution predicted by Rayleigh and Jeans. These curves are shown in Figure 16. The two are essentially identical for low frequencies (and therefore for long waves) but, where the real distribution drops for high enough frequencies, the Rayleigh–Jeans distribution goes off to infinity. This is because the number of allowed wave forms per unit of volume, per unit of frequency, increases rapidly with frequency, and the Rayleigh–Jeans picture says both that equal amounts of energy go into every allowed wave, and that these amounts of energy may be arbitrarily small.

That is to say, the Rayleigh-Jeans model attributes an infinite energy storage capacity or heat capacity to the high-frequency part of the radiation field, regardless of the temperature, because there is no limit to the number of oscillators at the high frequency end of the scale. Clearly, this is not a physically acceptable result; the radiation field cannot store an infinite amount of energy per unit volume in a system in equilibrium at any finite temperature. There are no infinite amounts of energy in the universe, as far as we know. Resolving this paradox is the second part of the puzzle that leads us to quantum theory.

The solid curves can be interpreted either as the displacement of a rope or the force on an electron from an electromagnetic field of force.

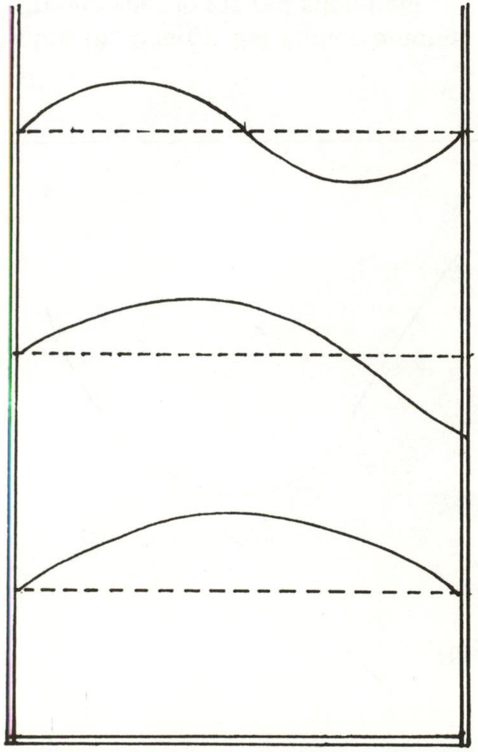

Allowed:
no wave amplitude at either wall.

Forbidden: nonzero amplitude at the right-hand boundary.

Allowed:
no wave amplitude at either wall.

Figure 15. Allowed and forbidden waves in a box

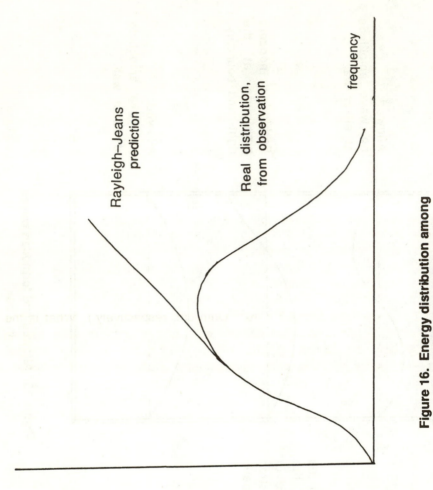

Rayleigh–Jeans prediction

Real distribution, from observation

frequency

Energy per unit volume, per unit of frequency

Figure 16. Energy distribution among frequencies in the radiation field.

Problems

1. How much energy, in joules, is stored in a 12 volt lead storage battery rated by the manufacturer as "400 ampere hours"?

2. Estimate the mass of a lead storage battery such as those used in automobiles. (The density of lead is about 11 gm/cm^3, the density of the battery acid and the hard rubber are about 2 gm/cm^3.) From the energy stored and your estimate of the mass, evaluate the energy stored per unit mass, a crucial quantity if the storage battery is to be a useful source of automotive energy.

3. Which is heavier, a meter length of copper wire or a meter length of aluminum wire, if both have the same resistance? (The density of copper is 9.0 gm/cm^3; the density of aluminum is 2.7 gm/cm^3. *Resistivities* of copper and aluminum are 1.7 × 10^{-6} Ohm cm and 2.6 × 10^{-6} Ohm cm, respectively.) What is the ratio of the prices of copper wire and aluminum wire having equal resistance, if the price of copper is $1.17/lb (or $2.58/kg) and the price of aluminum is $0.69/lb (or $1.52/kg), as stated in the *Wall Street Journal* of March 1, 1991?

4. What is the role of the frictional or retarding force in Ohm's Law? Indicate what characteristic of frictional forces is responsible for the existence of a terminal velocity for particles moving in force fields, and how terminal velocities are related to Ohm's Law. How would charges move if there were no friction? What role does the friction caused by air play in the behavior of sky divers? What frictional forces are important to skiers, especially if they are racing?

9. DILEMMAS OF ENERGY AND THE MICROSTRUCTURE OF MATTER

We have seen two dilemmas deriving from attempts to reconcile concepts of microscopic structure of matter and radiation with macroscopic concepts of heat and thermodynamics. One is the disparity between the real decreasing values of the heat capacity of solids at low temperatures, and the constant value predicted by our model of vibrating atoms. The other is the "ultraviolet catastrophe" of the Rayleigh–Jeans model for the capacity of the radiation field to store energy. Two other dilemmas also confronted physical scientists in 1900; the resolution of all four led us to the picture of energy and the structure of matter that we now accept. One of the remaining two is the problem of the photoelectric effect; the other is the problem of the stability of atoms and the kind of radiation that atoms can emit.

The photoelectric effect is a phenomenon first recognized by Heinrich Hertz when he produced radio waves, and then explored at the end of the 19th century, especially by Wilhelm Hallwachs and Philip Lenard. The phenomenon is best observed when visible or ultraviolet light illuminates a surface in a vacuum; the process is illustrated in Figure 17. In fact, one can see the process under some conditions even in air. Light, particularly ultraviolet light, incident on a surface, causes the flow of a current; Lenard showed that the current is due to the ejection of negative charges which we now know to be electrons from an irradiated surface.

Traditional ideas of light as a form of energy would lead one to expect the energy of the ejected electrons to be related to the energy contained in the light beam and therefore to the intensity of the light; one might even expect a direct proportionality relation between these two quantities. One might also expect the number of ejected electrons to be related to the frequency of the light, since the ability of any material to absorb radiation depends on the frequency of the light.

87

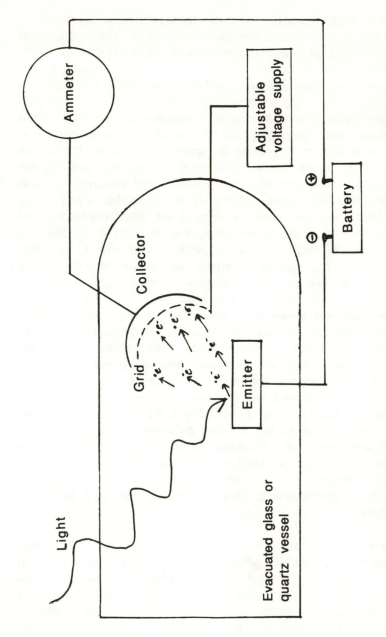

Figure 17. Apparatus for observing the photoelectric effect. Light from the source, of essentially a single frequency, strikes the emitter producing electrons that are then attracted to the collector. The grid is interposed to measure the energy of the electrons; if the grid is put at just high enough negative potential, it can stop the negative electrons from reaching the collector.

In reality, what is observed does not conform in the least to these expectations. Just the reverse: quite simply, the energies of the electrons emitted are proportional to the *frequency* of the light, and the number of electrons, the *photocurrent*, is proportional to the intensity of the light. How do we explain this puzzle?

The light gives a total energy E to each electron; if the light has a single color or frequency, all the electrons have the same energy. Some of this energy must be used to overcome the forces that keep the electron inside the metal; this amount is called W, the work function; this W is the amount of work each electron must do to escape the binding forces holding it in the solid. Any extra energy, E – W, appears as kinetic energy in the freed electron. This kinetic energy can be made to do work; the electron does this if it runs "up a potential hill" – – that is, if it encounters a repulsive force due to another negative charge, which changes its kinetic energy into potential energy. If, for example, the grid in Figure 17 is charged negatively, the electrons must slow down as they approach the grid. If the electrons are sufficiently energetic (i.e., if the repulsive force is not strong), then the electrons will be slowed down as they approach the grid but will nevertheless pass through. If, on the other hand, the grid is charged negatively enough, it will present an insurmountable "hill" and the electrons will never pass. By varying the voltage until the electrons *just* get through, one can measure the kinetic energy carried by the electrons when they leave the emitter. In this way, Lenard showed the relation between the energy of the emitted electrons and the frequency of the incident light. The puzzle this experiment presented was, as we have indicated, that the variables seemed to be wrongly connected to each other, according to the expectations people had about light and matter in the 19th century.

The fourth and last problem, that of the structure and radiative properties of atoms, seems far grander but is really not much more fundamental than that of the photoelectric effect. If one takes classical physics in a naive way to explain the structure of atoms and molecules in terms of electrons, protons and neutrons, then the picture seems clear: a stable atom exists because there are stable orbits for an object found in a potential that varies as $-1/R$, or that has an attractive force proportional to $1/R^2$.

This model is unfortunately incomplete, even in the context of 19th century science. In the complete, classical picture, a charged particle that moves under the influence of a force must radiate energy. If the charge is an electron in an antenna, then the radiant energy may transmit a radio or television broadcast. This is the example we discussed previously, a known and widely exploited process. If an electron were bound to a nucleus and obeyed the laws of classical physics, then that electron would slowly lose all its energy as it moved in the force field of its nucleus. The attractive force of the nucleus would continuously change the direction of motion keeping it in an orbit, and therefore would cause it to radiate energy continuously. As the electron's total energy decreased, it would spiral into the nucleus. Moreover the electron would move ever faster as it approached the nucleus, and therefore would radiate at progressively higher frequencies. Each atom would emit a continuous spectrum of frequencies, culminating in collapse of the electrons into the nucleus. By our own epoch, atoms would have ceased to exist.

In reality, atoms do not collapse and do not normally emit continuous spectra, containing all frequencies of light. They emit characteristic **spectral lines** – – light of very precise, specific wavelengths and frequencies characteristic of the specific kind of atom. They have well characterized states of lowest energy, called *ground states*, with electrons well established in stable states outside the nuclei. Reconciling this fact with the theory of the microstructure of matter was perhaps the most challenging task facing physical scientists at the beginning of the 20th century.

The problems of the Rayleigh–Jeans description of radiation and of the photoelectric effect were two of the touchstones from which sprang the first modern theory of the quantum nature of energy. Max Planck showed in 1901 that he could obtain an entirely valid representation of the observed energy distribution of Figure 16 by making a simple but revolutionary supposition. He postulated that the energy that could be stored by an oscillator with natural frequency v could only be integral multiples of a basic energy unit *directly proportional to the oscillator's characteristic frequency v.* (Remember that those oscillators of Rayleigh and Jeans are only conjectures, necessary components of their model.) We call each of these units of energy a **quantum** of energy. The constant of proportionality between energy and frequency is universally called h or Planck's constant, and has the approximate value 6×10^{-34} joule sec. We write the energy for one quantum,

$$E = h\nu , \qquad\qquad (37)$$

and for a state with n quanta,

$$E = nh\nu , \qquad\qquad (38)$$

for the energy in any one of Rayleigh and Jeans' oscillators of the electromagnetic field. Planck's condition forces the distribution curve of Figure 16 to drop at high energies because at any finite temperature, there is simply not enough energy per unit volume to make the large quanta required to excite the high-energy oscillators.

The photoelectric effect was interpreted very soon thereafter by Albert Einstein, who showed that the results observed by Lenard were just what should occur provided each electron absorbs the energy of a single quantum of radiation when it is excited and set free. The energy of the quantum serves two functions, to overcome the binding forces that hold the electron inside the metal and to give it kinetic energy to escape. Hence the two dilemmas of the radiation field were explained − − but only by adopting a concept of the nature of energy that seemed to contradict all that traditional physical science had built during the 18th and 19th centuries.

The first step toward resolving the dilemma of atomic stability was achieved by Niels Bohr and Arnold Sommerfeld, in 1913. Bohr proposed, in what seemed a very arbitrary way, that atoms exist with their electrons in stable orbits, and that only discrete orbits, corresponding to specific quantized amounts of energy or angular momentum, are permissible. The atoms in this model are prohibited from radiating or absorbing any energies except those corresponding to the differences between the energies of a final orbit and the initial orbit.

This model, arbitrary and unsatisfying as it was, achieved quantitative replication of the spectrum of atomic hydrogen and, with small empirical adjustments, of the alkali atoms as well. Clearly, there had to be something correct about the model.

Bohr's model for the hydrogen atom – – one electron and one proton – – can be described in a very simple and straightforward way, if we restrict ourselves to circular orbits and neglect the more complicated elliptical orbits. We also simplify the math a little by assuming that the proton's mass is infinite compared with that of the electron. The model requires that:

1.   the energy of the negatively charged electron consists of kinetic energy and potential energy due to the Coulomb attraction of the proton;

2.   the electron's orbit is stable because the repulsive centrifugal and attractive centripetal forces just balance; and

3.   the angular momentum of the electron is restricted to integral multiples of Planck's constant divided by 2π.

In terms of equations, these three sentences become, respectively,

$$1. \quad E = mv^2/2 - e^2/R \quad \text{(definition of energy)}, \tag{39}$$

$$2. \quad mv^2/R = e^2/R^2 \quad \text{(centrifugal force equals attractive force)}, \tag{40}$$

$$\text{and } 3. \quad mvR = nh/2\pi , \; n=1, 2, .. \text{ (quantization of angular momentum)} \tag{41}$$

where m is the electron's mass, v is its velocity, e is its charge, and

R is the radius of its orbit.

From Eq. **(40)**, we conclude that stability requires that $mv^2 = e^2/R$ or that the negative of the potential energy be fixed at a value of twice the kinetic energy. Hence the first equation, **(39)**, can be rewritten as $E = -mv^2/2$ or $= -e^2/2R$, and the third equation, **(41)**, can be rewritten in terms of v or R alone; if we write **(41)** as $m^2v^2R^2 = n^2h^2/(2\pi)^2$, then **(41)** also becomes $me^2R = n^2h^2/(2\pi)^2$ or

$$R = n^2(h/2\pi)^2(1/me^2) , \qquad (42)$$

which says that R, the radius of the orbit, is restricted to values proportional to the square of the integer n. Everything else on the right side of **(42)** is a natural constant. If **(42)** is reexpressed in terms of velocity, we obtain, from $mv^2 = e^2/R$,

$$mv^2 = e^2[n^2(h/2\pi)^2(1/me^2)]^{-1}$$
$$= (e^4m/n^2)(2\pi/h)^2 , \qquad (43)$$

so that the velocity

$$v = e^2(2\pi/h)(1/n) , \qquad (44)$$

is just inversely proportional to the integer n. Lastly, the energy itself becomes

$$E = -(e^4m/2)(2\pi/h)(1/n^2) . \qquad (45)$$

This expression is sometimes written as

$$E = -R_\infty/n^2 , \qquad (46)$$

where R_∞ is the *Rydberg* constant for a nucleus of infinite mass. The constant is named for the spectroscopist who showed that the *experimental* values of the frequencies of the lines in the spectrum of the hydrogen atom all fit the expression for their frequencies v:

$$\nu = R_H(1/n_i^2 - 1/n_f^2) \qquad (47)$$

where n_i and n_f are two integers. (The constant R_H, the Rydberg constant for hydrogen, is slightly different from R_∞, in a way, unimportant here, that can be computed precisely if one takes into account the finite mass of the proton.) In other words, every line of the spectrum of atomic hydrogen, from the far ultraviolet into the radiofrequency region, can be indexed by two integers, and these two integers are also sufficient to fix the frequency of the line.

The Bohr model says that one needs two integers to fix the frequency of a spectral line because radiation is emitted when an electron falls from an orbit of high energy to an orbit of lower energy, and each orbit is characterized by a single integer or quantum number n. The characterization of a spectral line requires specifying the quantum numbers of both the final state and the initial state. If the final state has energy $-R_H/n_f^2$, and the initial state has energy $-R_H/n_i^2$, then the change in energy of the electron is

$$E_{final} - E_{initial} = (-R_H/n_f^2) - (-R_H/n_i^2)$$

$$= R_H(1/n_i^2 - 1/n_f^2) \qquad (48)$$

and the energy radiated as a quantum of light must be equal to the energy *lost* by the electron:

$$E_{light} = E_{initial}(electron) - E_{final}(electron)$$

$$= R_H(1/n_f^2 - 1/n_i^2) \qquad (49)$$

hence the frequency of light emitted is $(R_H/h)(1/n_f^2 - 1/n_i^2)$

The model arbitrarily requires that n_j (that is, n_i or n_f) may be no smaller than 1, so that the electron is prohibited from falling into the nucleus. (If n_f could be zero, what would be the corresponding R?) However, n may be arbitrarily large, corresponding to arbitrarily large orbits, still nevertheless quantized in size and velocity. The higher is n_j, the larger is the radius and the smaller is the velocity. Interstellar hydrogen atoms have been found with the quantum number n changing from n_i = 250 to n_f = 249; these transitions were observed when the emitted radiation was detected by astronomers.

Just as energy is emitted by an atom when it loses internal energy – – that is, when the quantum number *n* changes from a larger integral value to a smaller value – – energy must be absorbed by an atom when the quantum number *n* increases in value. In effect, an electron moving from a small Bohr orbit with a high characteristic velocity to a larger orbit with a lower velocity must absorb energy from some external source. Light energy is one of the best sources, provided that quanta of the appropriate energies are present. Atoms in the outer layers of the sun are bathed in a continuous spectrum of light quanta. We can detect the presence of certain kinds of atoms because they *remove* light of specific frequencies from the solar spectrum, leaving dark lines called *Fraunhofer lines*. These lines appear at frequencies where there is very little energy getting from the sun to us. These **absorption lines** are just as characteristic as the emission lines, and occur at the same frequencies, of course. Absorption lines can be observed when atoms are colder than the radiation field in which they are bathed and consequently absorb energy from the field in order to come to thermal equilibrium with that field. Emission lines are observed when atoms are hotter than the radiation field around them, so lose energy to the field to "try to warm up the field".

Notice that the Bohr model predicts that as the quantum number *n* approaches infinity, the electron's orbit grows larger, but the electron remains bound, with an energy always less than zero. What of the levels with energy E greater than zero? These energies correspond to states in which the energy of the electron is everywhere greater than the potential energy. In such a state, the potential energy and the corresponding attractive force can always be overcome by the electron, which means that such an electron is free to escape to infinity. Clearly such a state does not correspond to a stable or bound orbit, yet is physically allowable. In fact, if an electron bound to a proton absorbs a quantum of sufficient energy, the electron is set free, in effect boiling off of the atom. The minimum energy required to set an electron free from an atom or molecule is the **ionization energy** of that atom or molecule. If more energy than the ionization energy is imparted to a bound electron, the excess remains with the electrons as kinetic energy. The ionization energy of an atom or a molecule is exactly the analogue of the work function W of a solid, which was discussed in connection with the photoelectric effect.

It is worthwhile noting at this point that the Bohr model places no quantum restrictions on electrons with energy greater than zero. The theory and the observations indicate that a free electron may take on any energy greater than zero, whether it is escaping from an atom or a bulk solid.

Problems

1. Which of the postulates of Bohr's model of the atom is responsible for the inference that the kinetic energy is half the negative of the potential energy? Which is responsible for the discrete nature of the possible stable orbits?

2. What is the frequency of the light emitted corresponding to a transition of a hydrogen atom from a quantum state with $n_i = 250$ to a state with $n_f = 249$? In what spectral region (ultraviolet, visible, infrared, microwave, radio or ultralong wave) should one look to observe this transition? What equipment would one use? Why might one want to observe such transitions?

3. How would the derivation of Bohr's equations differ from the form given in this Chapter if the single electron (charge $-e$) were in orbit around a nucleus with a charge $+Ze$, Z being an arbitrary integer?

4. Using the Bohr model of the hydrogen atom, derive an expression for the time $\tau(n)$ required for the electron to complete a single orbit with principal quantum number n. Express this first in terms of natural constants; then express the time as $\tau(n)$ = (numerical constant) × (n to some power). What is the time $\tau(n)$ for an orbit with $n = 150$? (Hydrogen atoms with $n = 150$ have been seen in interstellar space.)

5. A powerful ultraviolet laser can produce about 10^{17} light quanta in one flash, to generate radiation with a wavelength of approximately 300 nm (1 nm is 10^{-9} m). The duration of the flash is 10^{-8} sec. Approximately how many joules of light energy are produced in such a flash? If the light is emitted at a constant intensity during the 10^{-8} sec, what power (in watts) is the laser producing during this interval?

6. How much energy is required to remove an electron from its lowest bound state ($n = 1$) in a hydrogen atom to its free state of lowest energy ($E = 0$ or $n \to \infty$)? To what temperature T would this energy correspond, in the sense that we identify the energy as $k_B T$?

7. Describe how the dependence of the photoelectric effect on light frequency and light intensity implies that matter absorbs electromagnetic energy in discrete quanta, rather than in arbitrary quantities.

10. WAVE PROPERTIES OF MATTER

The next major advance beyond Bohr's model in our understanding of the microstructure of matter and of how matter contains energy came in several dramatic steps during the 1920's. Louis deBroglie proposed that if particulate character were ascribed to light, in the form of quanta, then one should consider that matter might have wavelike properties. In particular, deBroglie proposed that the momentum of a particle could be associated inversely with a wavelength, and that the proportionality constant be Planck's constant, h:

$$(\text{momentum}) \times (\text{wavelength}) = h$$

or, calling momentum p and wavelength λ,

$$p = h/\lambda \ . \tag{50}$$

If deBroglie's relation is to be consistent with the accepted energy of a free particle,

$$E = mv^2/2 \tag{51}$$

$$= p^2/2m \quad (\text{since the momentum } p = mv),$$

and with Einstein's relation between energy and frequency,

$$E = h\nu \tag{52}$$

then there must be a relation between the wavelength λ and the frequency ν of a matter wave. This relation has the form

$$h\nu = (1/2m)(h^2/\lambda^2)$$

$$\nu = h/2m\lambda^2 \ , \tag{53}$$

which is quite different from the expression connecting frequency and wavelength for light:

$$\nu = c/\lambda \ . \tag{54}$$

Clearly, a matter wave is not the same kind of beast as a light wave.

The wave picture introduced by deBroglie was made into a full mathematical treatment and developed for realistic problems by Erwin Schroedinger. Schroedinger developed a wave theory having some historical relationship to the theory of Maxwell for light waves. The dispersion relation – – the connection between wavelength and frequency – – dictates much of what form the equation must have. Schroedinger set the condition that the states of systems that correspond to particles in stable orbits, or to bound systems, must be represented by *standing waves*, in contrast to *running waves*.

An example of a running wave is the wave that travels out a once-shaken rope, or that is produced on the water's surface by a stone dropped into a still pond. A standing wave is one generated by twirling a jump-rope, or by blowing a flute. The fundamental difference between running waves and standing waves is that running waves may travel as far as the system allows, and can move (propagate) indefinitely far in free space. Standing waves require, for their existence, constraint by some kind of boundary, such as the rim of a drumhead, the ends of an organ pipe, or the hands that hold a jump-rope.

The constraints that permit a system to exhibit standing waves also prevent that system from having standing waves of arbitrary shape, wavelength or frequency. Figure 15 showed examples of two allowed standing waves in a simple box, and one wave that cannot exist in that box as a standing wave. In general, only a very limited, discrete set of standing waves are even possible for any real system. This is why a bugler cannot play full scales; the valve trumpet, with its length adjustable by means of lengths of tubes and valves, is really several sizes of bugle combined into a device with one mouthpiece and one bell. The variable length increases the number of tones a trumpet can sustain.

The restriction of waves representing bound systems to the limited, discrete set of standing waves is precisely the origin of quantization in the Schroedinger wave-mechanical picture. Each wave-state corresponds to a particular energy and is the counterpart of a stable orbit. However, one should certainly not identify the classical picture of particles in orbits with the standing waves of quantum mechanics. Occasionally, attempts are made to find a classical particle-like basis for the wave models, but none has been successful in any general or fundamental way.

At the same time Schroedinger was developing the wave picture, Werner Heisenberg was among the group of physicists trying to find a new basis for physics that answered the criticisms of the logical positivists who argued that science should deal only with observables. Although this goal has never been totally achieved, Heisenberg did derive a form of mechanics, superficially quite different from that of Schroedinger, but exactly as powerful and accurate as wave mechanics. Both approaches could be used to solve the same problems and both gave the same answers. P.A.M. Dirac showed that the two methods were actually different representations of the same underlying physical picture, in which a system is entirely characterized by the forms of its forces, its boundary constraints, and a finite set of quantum numbers. It is neither necessary nor possible, according to Dirac's general formulation of quantum mechanics, to describe a system with all the variables of position and momentum that classical mechanics requires. Instead of a single picture requiring knowledge of positions and momenta (or velocities), quantum mechanics offers a variety of descriptions of any system. Any desired description in terms of positions can be obtained from a description in terms of momenta. Any description is based on what one calls a *complete set of compatible variables*.

The notion of compatibility arose in connection with the interpretation of the meaning of the matter wave. Max Born brought forth, and Niels Bohr and others helped develop the concept that the *intensity* of a matter wave at any point (the absolute value of the square of the amplitude of the displacement) should be interpreted as the density of *probability* of finding the corresponding particle of matter in the vicinity of that point. But if one tries to locate a particle, one's apparatus interacts with the matter wave of that particle, diffracting the wave and imparting additional curvature to that wave. This shortens the wavelength locally, and disturbs the momentum of the particle. Each time we try to pinch the particle to locate it, it slips away from our fingers because we interfere with its wave form.

The inability to locate the position of a particle because of the unknown amount of momentum we inadvertently give the particle during our measurement is one facet of the *Uncertainty Principle*, first formulated by Heisenberg. In more general terms, Heisenberg showed that natural observables come in complementary pairs; measurement of one member of a pair necessarily introduces some uncertainty in the value assumed by the other quantity. Position and momentum form one pair; energy and time (duration of the interval required to conduct the measurement) are another; angular momentum and angle are still another. These pairs all have in common that their products have the same dimensions as Planck's constant. In fact, Heisenberg's Uncertainly Principle requires that the product of the uncertainties of the two variables of a pair be greater than Planck's constant h divided by 4π.

Clearly, the two variables in any complementary pair do not correspond to *compatible* variables because measurement of one disturbs the value of the other. A set of compatible variables is a set such as positions along the three directions of space, measurements of which do not interfere with each other. Dirac's formulation of quantum mechanics led to the recognition that compatible variables come only in discrete, *finite* sets. The different choices of descriptions correspond to different choices among the sets of compatible variables.

One implication of the Uncertainty Principle is of particular concern with regard to our study of energy. This comes from the complementarity between energy and time. If a system such as an excited atom is capable of losing energy in some way such as radiation, then the length of the interval during which the atom contains its excess energy is uncertain. This uncertain time interval, Δt, is related to the uncertainty we have in how much energy the atom actually contains. If the uncertainty in energy is ΔE, then the Uncertainty Principle tells us that

$$(\Delta E)(\Delta t) \ > \ h/4\pi \ . \tag{55}$$

This uncertainty in the time during which the atom contains the excess energy before radiating is often called the *natural radiative lifetime* of an atom in a particular state. The uncertainty in the energy of that state increases as the natural lifetime of the state gets shorter, according to Eq. (55). This is in fact what is observed.

The wave structure of matter is most easily understood by considering the wavelengths of the elementary particles, electrons and protons, as they depend on the energy of the particle. A particle with a fixed kinetic energy moving along some direction has a fixed momentum in that direction, which fixes the wavelength of the wave corresponding to that particle. A proton moving at a velocity of about 2×10^4 meters/sec (corresponding to a mean kinetic energy of $^3/_2 \, k_B T$ near that appropriate to room temperature) is a good first example. The momentum, mv, of the proton, is

$$1.6 \times 10^{-27} \text{ kg (i.e. } 1/N_A) \times 2 \times 10^4 \text{ meters/sec.}$$

Hence the wavelength, according to deBroglie's relation, is

$$\lambda = h/p$$

$$= 6 \times 10^{-27} \text{ erg-cm}/3.2 \times 10^{-18} \text{ gm cm/sec}$$

or

$$6 \times 10^{-36} \text{ joule-meters}/3.2 \times 10^{-23} \text{ kg-m/sec}$$

$$\equiv 2 \times 10^{-9} \text{ cm or } 2 \times 10^{-11} \text{ meter or } 0.2\text{Å} \ , \tag{56}$$

somewhat smaller than a conventional atomic diameter, but large compared with our usual ideas of the size of an atomic nucleus. If we increase the energy of the proton a hundredfold, then its velocity increases tenfold, and its wavelength decreases by a factor of ten.

An electron has a mass about 1/1800 that of the proton.* Let us estimate the wavelength of a "thermal" electron and compare this with the wavelength of the proton whose kinetic energy is "thermal", i.e. equivalent to $^3/_2 k_B T$ when T is room temperature. The electron and proton, both being thermal, have the same kinetic energy. Therefore, in a mathematical sentence,

$$m_{electron} v^2_{electron}/2 = m_{proton} v^2_{proton}/2 \qquad (57)$$

or, if p = mv = momentum,

$$p^2_{electron}/2m_{electron} = p^2_{proton}/2m_{proton} \qquad (58)$$

so that

$$p_{electron} = p_{proton} \times \sqrt{(m_{electron}/m_{proton})}$$

$$= \sqrt{1/1800} \; p_{proton} \; . \qquad (59)$$

Hence

$$\lambda_{electron} = h/p_{electron} = (h/p_{proton})\sqrt{1800/1} \qquad (60)$$

and since we just evaluated p_{proton} as about 3×10^{-18} gm cm/sec,

$$\lambda_{electron} = (6 \times 10^{-27} \text{ erg-sec}/3 \times 10^{-18} \text{ gm cm/sec}) \times \sqrt{1800}$$

$$= 2 \times 10^{-9} \text{ cm} \times 42$$

$$= 8.4 \times 10^{-8} \text{ cm, or } 8.4\text{Å} \qquad (61)$$

Hence the thermal electron has a wavelength roughly 8 times the diameter of a typical atom.

---

* See M. Shamos, *Great Experiments in Physics*, Henry Holt, New York, 1959 Chapt 16.

How can it be that the proton is almost as big as an atom and the electron is so much bigger? The answer is, of course, that our examples thus far have been of very slow particles, with far less kinetic energy than that of an electron bound in an atom. In fact, we know that the energy required to remove an electron from an atom is typically 10^{-18} joule or more. This means that about 10^{-18} joule is necessary to overcome the potential "hill" that binds the electron to its parent nucleus. But if the potential energy is about 10^{-18} joule, then the electron should have a comparable amount of kinetic energy, on the average, to keep it from falling into the nucleus.* To achieve a stable state, the electron, like a planet moving around the sun, must have a precise balance between kinetic and potential energy. If the kinetic energy is too high for its potential, a planet will fly off into space; if the potential is too great an attraction to be balanced by the kinetic energy, the planet falls into the sun. The electron and nucleus have no direct analogue to that collapse, but the conditions for stability are parallel to those of the solar system.

The wavelength of an electron bound to a proton, having a binding energy of about 10^{-18} joule, can be derived this way:

_____

* For any truly bound system such as a pair of particles held by attractive electrical forces or a particle bound on a spring, there is always a unique relationship between the average kinetic energy and the average potential energy. The relation depends on the particular form of the attractive potential. For example, a particle bound by an ideal spring has, on average, exactly as much kinetic energy as potential energy. A particle bound by electrostatic attraction has a potential energy that is negative. (The energy of two infinitely-separated oppositely-charged particles with no kinetic energy is usually taken as zero, and the potential energy is negative for any finite separation. This is the way we set the energy scale in Chapter 9.) The kinetic energy is of course positive. For a stable system bound by $1/R^2$ attractive forces (or $1/R$ attractive potentials, which amounts to the same thing) whether electrical or gravitational, the average kinetic energy is half the negative of the average potential energy, as we saw in Chapter 9:

$$(KE)_{average} = -\tfrac{1}{2}(PE)_{average}$$

Hence, the deeper in the potential a particle is, the faster it must go to maintain its orbit. The orbital periods of the planets are clear examples of this phenomenon.

_____

$$mv^2/2 = p^2/2m = 10^{-18} \text{ joule,}$$

so $$p^2 = 2 \times 10^{-30} \times 10^{-18} \text{ kg-m}^2/\text{sec}^2 \qquad (62)$$

and $$p = 1.4 \times 10^{-24} \text{ kg-m/sec} \qquad (63)$$

and $$\lambda = (10^{-34} \text{ kg-m}^2/\text{sec})/(1.4 \times 10^{-24} \text{ kg-m/sec})$$

$$\equiv 0.7 \times 10^{-10} \text{ m or } 0.7\text{Å} \qquad (64)$$

Thus, the average wavelength for an electron bound to a proton, as in a hydrogen atom in its lowest stable state, is just about the size of the atom. In fact, it is *precisely the spatial distribution of the electron waves that determine the physical dimensions of atoms and molecules.*

The waves corresponding to stable orbits of electrons in atoms increase in size as the energy of the electron approaches zero from below. In fact, the position of the maximum amplitude of a spherical wave in a one-electron atom falls at the position of the radius of the corresponding circular orbit in the Bohr atom. The radius of a circular Bohr orbit, as Eq. **(39)** shows, is proportional to n^2, where *n* is a positive integer. The proportionality constant for the hydrogen atom, $h^2/4\pi^2 e^2 m_{\text{electron}}$, is approximately 0.53Å, and is called the Bohr radius, a_o. Hence the radius of a hydrogen atom as measured by the classical Bohr model or by the position of the maximum amplitude of the electron wave, as described by wave mechanics, is given by the square of a **quantum number** *n*, times the Bohr radius a_o:

$$\text{"radius"} = n^2 a_o \qquad (65)$$

The larger the wave, the further out it spreads into regions of shallow potential. But if an electron spends most of its time in regions of shallow potential or weak attractive force, then its kinetic energy must be small, on average, which in turn means that its momentum must be small. If the momentum is small, then the corresponding wavelengths must be large. Thus, with increasing quantum number *n* and increasing radius, the bound electron takes on an increasing average wavelength.

Just as the standing waves on a drumhead may take on various shapes, the standing electron waves corresponding to stable states of an electron may take on various shapes. It is easiest to characterize the shapes of standing waves by examining the curves or surfaces of zero displacement, the so-called *nodal* surfaces. On a vibrating drumhead, essentially a two-dimensional wave pattern is established; the nodes of the standing waves on this surface are of two kinds, circles and diameter lines. The lowest vibrational mode of a drumhead, called the fundamental, has the drumhead all vibrating at once and in the same direction. In the fundamental mode, the only curve of zero displacement is the circle at the rim, where the drumhead is tied down. The next modes each has one nodal curve or line, and there are two types, the one with a single circular node, and those with one diameter acting as a node. There are just two independent but physically equivalent modes of this latter sort, corresponding to the two independent directions, x and y, in a plane. A mode involving any other single nodal diameter can be represented as a combination of the x-like and y-like modes. These and some of the higher modes of vibration of a drumhead are shown in Figure 18.

The standing waves of an electron wave in three dimensions bear a close relationship to the standing waves of a drumhead in two dimensions. The lowest state is a nodeless, spherically-symmetric distribution. Because of the infinite depth of a potential that varies as R^{-1}, the amplitude collects and concentrates around $R = 0$, to form a cusp there, for the state of lowest energy. Standing waves corresponding to states of higher energy may have spherical nodes, planar nodes or conical nodes. The energy of an electron's state increases with the number of nodes in the wave; the more nodes, the nearer is the energy of the state to zero, and the less the energy required to bring the electron's energy up to zero and thereby to detach it. If the electron has no angular momentum, its wave can concentrate somewhere near the origin, as in the state of lowest energy. If the angular momentum of the electron is not zero, centrifugal force makes it impossible for the wave to collect around the origin.

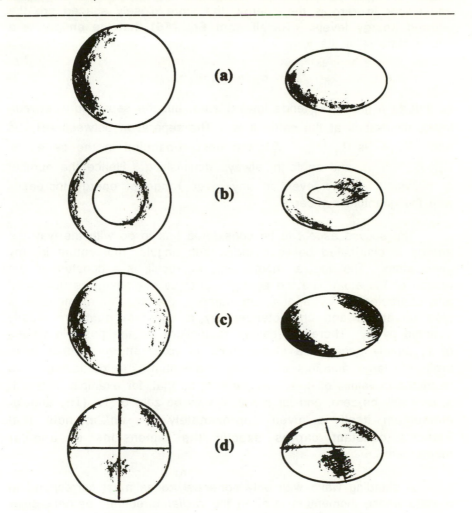

Figure 18. Four modes of vibration of a drumhead: a) fundamental; b) one circular node; c) one linear node; and d) two linear nodes.

There are two very important aspects to the energy levels of electrons bound to atoms that we must recognize. First, the higher and closer to zero is the energy, the more closely spaced are the allowed energy levels. Recall from Eq. **(45)** that the energy of a bound state is of the form

$$E_n = R_H/n^2 ,$$

so that as **n** grows, E_n tends toward zero, but the separation between levels diminishes at the same time. The separation between $-R_H/3^2$ and $-R_H/4^2$ is $R_H(^1/_{16} - ^1/_9)$; the next separation in the series is $R_H(^1/_{25} - ^1/_{16})$, and so forth, always diminishing. Hence the number of allowable energy levels in any given range of energy increases with the quantum number **n**.

The second aspect to be considered has to do with the way the energy is distributed between radial and angular momentum in any given state. The angular momentum, we recall, is a constant of the motion of the atom, like the energy. Just as the energy is quantized and characterized by a quantum number, the angular momentum is also quantized and characterized by a quantum number, usually denoted by ℓ. There are severe restrictions on the possible values of ℓ, simply because a stable state of fixed energy cannot have arbitrarily large amounts of angular momentum. In particular, ℓ is restricted to values of 0, 1, . . . , **n** − 1, so that, for example, if **n** = 1, ℓ can only be zero, and for **n** = 2, ℓ may be zero or 1. (The angular momentum itself is given approximately by $\ell h/2\pi$; note that Planck's constant carries exactly the dimensions of angular momentum.)

A standing wave with only spherical nodes must correspond to a state whose momentum is all in the *radial* direction, perpendicular to the wave fronts, so that such a wave corresponds to a state with zero angular momentum. The number of *angular* nodes, conical or planar, determines the number of units of angular momentum. The lowest electron waves with **n** = 2, ℓ = 1, in a hydrogen atom, can be thought of as having one planar node perpendicular to any one of the axes x, y or z. These three are all physically equivalent, but independent because of the three independent dimensions of space. In effect, we are seeing a property of the symmetry of space. Note that there are, consequently, 4 kinds of standing waves for the energy state with n = 2; one has a single spherical node, and the others have planar nodes.

Now we come to the crux of this part of our discussion of waves. How many independent standing waves are there for the energy level characterized by the arbitrary integer n? The answer is that there are n^2 such levels. This is perhaps rather startling; it means that the number of possible states (which means counting each independent standing wave as one possible state) is n^2, for the energy level characterized by the quantum number n. Hence, not only does the spacing between allowed energies decrease with n, but the number of states at each n increases. The net result is that the total number of levels contained in a chosen narrow band of energy increases very rapidly, as n^5 in fact. This in turn implies that, although it is very easy to pick out individual quantum states when n is low, it rapidly becomes very difficult to sort out specific states when the quantum number n gets large. Moreover it is not only hard for us to distinguish the levels; it becomes hard for the electron also.

We can imagine the energy levels of a hydrogen atom as a succession of balconies in a theater, each one with a larger seating capacity than the one below it. The states correspond to the seats in the theater. Only the orchestra has a single seat; the first balcony has 4 seats, the second, 9; the third, 16; the fourth, 25; the fifth, 36; and so on. The electron is like the single spectator who comes in to this theater to sit down. If she goes up to the nineteenth balcony, she has a choice of 400 seats!

How do electrons fit into atoms more complex than hydrogen? Two simple principles govern this problem. The electrons go into the energy levels as though they, the "balconies", were shells and then into the states within the levels or shells. The states have small differences in energies within the shells, depending on the amounts of angular momentum. We can imagine each balcony of our metaphorical theater to be built with a sloping floor, so that not all seats in each balcony are on precisely the same level. However the differences of energies between shells or balconies are larger than those within a shell, generally.

Then we should expect that in the most stable state, the state of lowest total energy, all the electrons would reside in the shell of lowest energy. This brings us to the second of our principles, which prevents this: the **Exclusion Principle** of Wolfgang Pauli. This Principle says that no more than one electron may ever occupy a given seat. We must actually go a bit further in designing our theater; we must take into account one property of electrons that we have not yet recognized, the internal intrinsic angular momentum of the electron, which we call spin. There are two possible states of the internal spin for any electron, associated with the orientation of its spin angular momentum in space. This allows the presence of $2n^2$, rather than n^2 electrons in each shell. We must accept double the number of seats in the theater, to account for spin.

The helium atom contains two positive charges in its nucleus and therefore two electrons outside the nucleus. These fill the shell with $n = 1$. The next shell can hold 2 × 4 or 8 electrons; the addition of successive positive charges to the nucleus and, to keep electrical neutrality, of electrons outside the nucleus to fill the second shell, corresponds to building up the elements from lithium through neon. The third shell begins with sodium; when another eight electrons have been added after the ten of neon, one has built up an atom of argon. The next two, the nineteeth and twentieth electrons, go into the first row of the balcony with $n = 4$. However, the next 10, rather than going to the second row of the balcony labeled $n = 4$, prefer the back or highest-energy part of the balcony with $n = 3$; these are the places with $n = 3$ and $\ell = 2$. When these seats are filled, the next electron customers must go up to $n = 4$ and $\ell = 1$. To what elements do all these filling steps correspond? Consult the *Periodic Table of the Elements* to identify the steps.

1																	2
H 1.008																	He 4.003
3 Li 6.94	4 Be 9.01											5 B 10.81	6 C 12.011	7 N 14.01	8 O 16.00	9 F 19.00	10 Ne 20.18
11 Na 22.99	12 Mg 24.31											13 Al 26.98	14 Si 28.09	15 P 30.97	16 S 32.06	17 Cl 35.45	18 Ar 39.95
19 K 39.10	20 Ca 40.08	21 Sc 44.96	22 Ti 47.90	23 V 50.94	24 Cr 52.00	25 Mn 54.94	26 Fe 55.85	27 Co 58.93	28 Ni 58.71	29 Cu 63.55	30 Zn 65.37	31 Ga 69.72	32 Ge 72.59	33 As 74.92	34 Se 78.96	35 Br 79.90	36 Kr 83.80
37 Rb 85.47	38 Sr 87.62	39 Y 88.91	40 Zr 91.22	41 Nb 92.91	42 Mo 95.94	43 Tc 98.91	44 Ru 101.07	45 Rh 102.91	46 Pd 106.4	47 Ag 107.87	48 Cd 112.40	49 In 114.82	50 Sn 118.69	51 Sb 121.75	52 Te 127.60	53 I 126.90	54 Xe 131.30
55 Cs 132.91	56 Ba 137.34	57 La 138.91	72 Hf 178.49	73 Ta 180.95	74 W 183.85	75 Re 186.2	76 Os 190.2	77 Ir 192.2	78 Pt 195.09	79 Au 196.97	80 Hg 200.59	81 Tl 204.37	82 Pb 207.19	83 Bi 208.98	84 Po (209)	85 At (210)	86 Rn (222)
87 Fr (223)	88 Ra 226.03	89 Ac (227)	104 (Rf) (261)	105 (Ha) (262)	106 (263)												

Lanthanides

58 Ce 140.12	59 Pr 140.91	60 Nd 144.24	61 Pm (145)	62 Sm 150.35	63 Eu 151.96	64 Gd 157.25	65 Tb 158.93	66 Dy 162.50	67 Ho 164.93	68 Er 167.26	69 Tm 168.93	70 Yb 173.04	71 Lu 174.97

Actinides

90 Th 232.04	91 Pa (231)	92 U 238.03	93 Np (237)	94 Pu (244)	95 Am (243)	96 Cm (247)	97 Bk (249)	98 Cf (249)	99 Es (254)	100 Fm (257)	101 Md (258)	102 No (259)	103 Lr (260)

Numbers in parentheses: available radioactive isotope of longest half-life.

Figure 19. Periodic Table of the Elements

Problems

1. The photoelectric effect can be used to measure the binding energies that hold inner-shell electrons in gaseous atoms, as well as the work function W of electrons bound in solids. One only has to use photons of sufficiently high energy, and X-rays have such energy. Describe a hypothetical experiment similar to that of Figure 17, with which one might measure the binding energies of the electrons in different shells of some species of atom. Sketch the form in which you might expect to collect your data (in a graph, for example).

   a. Compute the deBroglie wavelengths for an electron moving at a speed of 6×10^6 m/sec (2% of the speed of light). Assume Planck's constant $h = 6 \times 10^{-34}$ joule-sec and the mass of the electron is 10^{-27} gm. Compute the wavelengths for an electron whose kinetic energy $= k_B T$ at room temperature, about 300°K. (First step: find the velocity of this electron.)

   b. Compute the deBroglie wavelength for a proton (mass = 2000 × mass of electron) moving at a speed of 6×10^6 m/sec, and for a proton with a kinetic energy of $k_B T$ at room temperature.

   c. Compute your own deBroglie wavelength when you are moving with a speed of about 0.6 m/sec.

   d. Look for a moment or two at the wavelengths you have just calculated. Compare them with other sizes, like atomic diameters.

2. The product of a frequency and a wavelength is necessarily a velocity, at least dimensionally. If one uses the Einstein relation $E = h\nu$ and the deBroglie relation $p\lambda = h$, one finds that for matter waves, $\nu\lambda$ is a constant for a given particle. What is the relation between the product $\nu\lambda$ and the *particle's velocity* (remember $p = mv$) for a matter wave? This product is not equal to the particle's velocity v.

11. WAVES, ENERGY LEVELS AND DENSITIES OF STATES

The physically allowed states of an electron bound to a proton are described by a set of standing wave states characterized by the number of nodes in the wave and by a quantized energy $E_n = -R_H/n^2$. The number of nodes is $n - 1$, and these may be radial or angular. In the hydrogen atom, all the states, n^2 in number, that correspond to a given value of n, have the same energy. We say, by way of definition, that the *degeneracy* of the nth level is n^2-fold. Moreover all the bound states of the hydrogen atom fit into this picture; none are omitted. States with energy $E > 0$ are states in which the electron may escape, rather than bound states.

Now let us look at the spacing between the energy levels. In the hydrogen atom, and in more complex atoms with a single outer electron as well, the allowed values of the energies of bound states form a sequence converging to $E = 0$ from below, and with the general form $E_n \propto -1/n^2$. Hence the levels get closer as the energy increases. As the discussion on pages 101–103 points out, the number of energy states, per (narrow) unit band of energy increases as n^5, a rapid increase indeed!

Not all systems exhibit such a decrease in their level spacing. A particle confined to move in a box in one dimension has allowed energies that correspond to the standing waves in a jump-rope with its ends held. These states have energies that also can be expressed in terms of a quantum number n, where again the number of nodes is $n - 1$:

$$E \text{ (particle in a 1-dimensional box)} \propto n^2 , \quad \textbf{(66)}$$

or, more precisely, if the box has length L, and the particle has mass m,

$$E = (\pi^2/L^2)(h/2\pi)^2(1/m) \, n^2 \qquad \textbf{(67)}$$

114

Thus the energy levels of a particle in a box in one dimension become more widely separated as the energy increases. Moreover, the levels of this system are all nondegenerate; there is only one state for each allowed energy.

A particle confined in a 3-dimensional cube, with sides all L in length has allowed energy levels whose spacings also increase, but these levels are characterized by three quantum numbers, one for each direction, rather than only one. For a 3-dimensional cubic box,

$$E = (\pi^2/L^2)(h/2\pi)^2(1/m)(n_x^2 + n_y^2 + n_z^2) \ . \qquad (68)$$

Because there are various ways to assign integral values to n_x, n_y and n_z to obtain the same values, ($1^2 + 2^2 + 1^2 = 2^2 + 1^2 + 1^2 = 1^2 + 1^2 + 2^2$, and more subtle cases, e.g. $5^2 + 5^2 + 1^2 = 7^2 + 1^2 + 1^2$). Almost all of the states above the lowest are degenerate, at least 3-fold and sometimes more.

Another example whose allowed energy values become more widely spaced as the energy increases is that of a rotating object such as a rotating molecule. As with the particle in a box, the levels take on values that are (approximately) proportional to the square of the integer usually designated as J. In this case, the degeneracy is a simple and very important one. The degeneracy of the level with quantum number J, whose energy is $E_J = const \times J^2$ is:*

$$g_J = \text{degeneracy of the } J^{th} \text{ rotational level} = 2J + 1 \qquad (69)$$

This number corresponds to $2J + 1$ different allowed spatial orientations for the axis of rotation. In other words, space itself takes on a quantized aspect.

---

* Actually, $E_J = J(J + 1)(h/2\pi)^2 \times$ (moment of inertia)$^{-1}$. The moment of inertia measures the leverage of all the masses of the particles. It is the sum over all masses of each mass times the square of its distance to the center of all the mass.

We have seen three systems in which the level spacing increases with energy (rotator, 1-dimensional particle in a box and 3-dimensional particle in a box), and one system (a single atomic electron) in which the level spacing decreases with increasing energy. In most systems we know, the degeneracy of the levels remains constant or increases with energy.

Is there any system with the same energy interval between each level and its neighbors above and below? The answer is yes, there is one, the quantum mechanical harmonic oscillator. The allowed energy levels of this system are spaced in equal intervals; the spacing is equal to Planck's constant, times the frequency ν of the corresponding classical oscillator:

$$\Delta E = h\nu \qquad (70)$$

between all pairs of levels. This is particularly interesting because it tells us that any force generating a more gently-rising potential than the parabola of the harmonic oscillator gives rise to level spacings that decrease with increasing energy. Potentials steeper than the parabola generate levels whose spacing increases with energy. These are sketched in Figure 20.

Figure 20 illustrates one other significant facet of the quantum view of matter. In any bound system, the lowest quantum state corresponds to a finite, non-zero energy. In both the harmonic well and the box, the lowest energy of the well is called zero, but the lowest quantum state has an energy above zero. This quantity is called the zero-point energy. The zero-point energy and the consequent delocalized distribution of positions and momenta (or velocities) for even the least energetic state of any system display vividly how nature forces the Uncertainty Principle onto us. Were the lowest allowed state of the harmonic oscillator to have zero energy, we would know both its momentum (zero) and its position (zero displacement) precisely. The requirement that the lowest state have some energy, actually $\frac{1}{2}h\nu$, ensures that the uncertainty principle is satisfied.

116

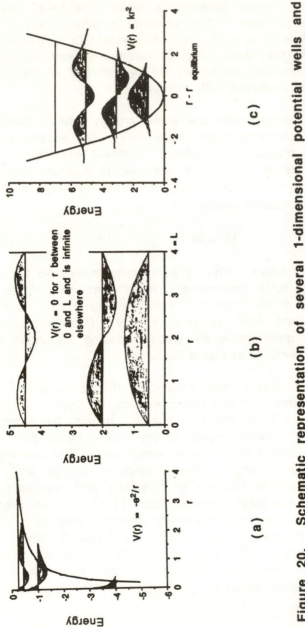

Figure 20. Schematic representation of several 1-dimensional potential wells and the energy levels supported in them. Energy is plotted on the vertical axes; at each allowed energy level, the amplitude of a standing-wave state is also shown as a displacement of the vertical axis, centered at the corresponding energy level.
(a) Coulomb potential well, converging level spacing; only 3 wave functions shown.
(b) Particle in a 1-dimensional box, diverging level spacing.
(c) Harmonic oscillator in 1-dimension, equal level spacing.

In Chapter 16 we shall return to the questions of level spacing and degeneracy in the context of the Second Law of Thermodynamics. However, to anticipate that discussion, it is worthwhile to point out the reason for introducing these ideas. A system tends to occupy all the quantum states available to it. So long as a system of *very* many atoms is hot enough, there will always be a few in highly excited states. The probability that atoms will be found in a particular energy level is proportional to the number of distinct quantum states having that energy, since presumably all the states of a given energy are equally probable if we can neglect the time required for transitions. But if the number of levels in any energy interval increases rapidly with energy, then so must the probability for finding an atom in any energy interval. It seems that this argument could lead us to the conclusion that all the matter in the universe should slowly accumulate energy from the energy-rich (albeit cold) radiation field in space and eventually dissociate.

More sophisticated and rigorous mathematical arguments can be used to derive this paradox, but the apparent conclusion remains, that the somewhat dismaying prospect of the total dissociation of all matter is inevitable, and probably should have happened already.

The paradox can be resolved, at least in large part, if the universe is finite. If this is the case, then for quantum states, with sufficiently high quantum numbers, the atomic electrons see the "walls" or boundaries of the universe far more than they see the nuclei to which they are supposedly bound. In this event, the level spacings eventually become essentially those of particles in boxes, with *increasing* intervals. With this restriction, it becomes possible to have bound particles that are stable.

Problems

1. What is the energy separating the levels with $n = 1$ and $n = 2$, of a particle weighing 1 atomic mass unit (1.6×10^{-27} kg), in a 1-dimensional box with $L = 1$ meter? $L = 10^{-6}$ meter? What is the separation of the levels $n = 2000$ and $n = 2001$ in each of these boxes?

2. If the rotational quantum levels for $J = 0$ and $J = 1$ are separated by an energy that corresponds to the frequency $v = 3 \times 10^9$ sec^{-1}, at what value J_o are the levels J_o and $J_o + 1$ separated by an energy approximately equal to kT at room temperature?

3. Explain why the particle in a 3-dimensional box has energy levels that are degenerate, while those for a 1-dimensional box are not. Consider, instead of the number of quantum states at a specific energy, the number of quantum states in a finite interval of energy. How does this number change with energy for the particle in a 3-dimensional box? For the 3-dimensional harmonic oscillator?

12. MOLECULES AND CHEMICAL BONDS:
ENERGY STORAGE IN MOLECULES

ELECTRONIC STRUCTURE OF MOLECULES: CHEMICAL BONDS

The shell structure of atoms is a result of the Coulomb force field, the relatively large mass of the nucleus compared with that of the electron, the Pauli Exclusion Principle, and the requirement that electrons be represented by standing waves. Molecules have similar shell structure, for just the same reasons. However the variety of molecules makes it difficult to describe molecular shell structure in as general a way as we used for atoms.

The innermost shells of electrons in molecules are close to their nuclei; consequently they are very much like the corresponding shells of isolated atoms. The higher-energy occupied shells, especially the uppermost one, two or three, vary considerably in shape and energy, depending on the atoms comprising the molecule. The details of the shapes of these shells need not concern us, but one general aspect of their shapes is particularly helpful for understanding the origins of chemical bonds.

Consider two positively-charged nuclei, A and B, with nuclear charges $+Z_A e$ and $+Z_B e$, separated by a distance R, as shown in Figure 21a,b. Now let us examine the effect of the force on these nuclei caused by a bit of negative charge $-q$. In particular, what is the effect on the separation of the nuclei of the attractive forces between this charge and the nuclei? Clearly, if the negative charge is in the region between the nuclei, where it attracts both nuclei toward itself, its effect is to shorten the distance R. Negative charge located between the nuclei therefore tends to strengthen and tighten a bond between A and B.

Suppose now, in contrast, that the negative charge is located outside the region between the nuclei; specifically that it is at position 2 in Figure 21a. Here, the attractive force of the charge $-q$ is stronger for B, the nearer nucleus, than for A, the more distant

120

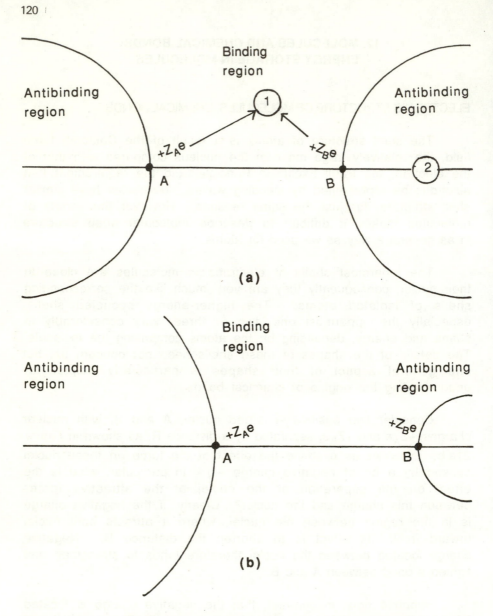

Figure 21. **Binding and antibinding regions of space.** (a) **Equal nuclear charge $Z_A = Z_B$.** (b) **Unequal nuclear charges $Z_A > Z_B$.**

nucleus, and both forces are in the same direction. The net effect of both attractions on R is to tend to *lengthen* the internuclear distance. When the negative charge is "beyond the nuclei", as in position 2 this is always its effect if the two nuclei have equal charges; for unequal nuclear charges, the negative charge pulls the nuclei apart if the negative charge is close enough to the nucleus with the larger positive charge. In these cases, the negative charge tends to <u>weaken</u> a bond between A and B.

Suppose the charge on B is smaller than the charge on A, and the negative charge $-q$ lies beyond nucleus B. Then we find that the negative charge tends to pull A and B apart if it is close to B, but acts to pull A and B together if it is sufficiently far from A and B. That is, if $-q$ is close to B, it pulls harder on B, but if it is far from A and B, it pulls harder on A. We can see this explicitly by writing

$$\text{Force on B} = -Z_B eq/r_B^2 \text{ ,}$$

and $$\text{Force on A} = -Z_A eq/(r_B + R)^2 \text{ .}$$

These quantitative expressions tell us in precise terms what our intuitions tell us more diffusely: when r_B is small, the denominator in the first expression is very small, and the force on B becomes correspondingly large, but the denominator in the second expression takes on a value only slightly larger than R^2, so that the force on A approaches $-Z_A e^2/R^2$. If r_B is much larger than R, then both denominators are very close in value, and both can be approximated as r_B^2. In this case, if $Z_A > Z_B$, then the force on A becomes larger than the force on B, and the nuclei are pulled together by the effects of the negative charge.

We have distinguished regions of space in which negative charge acts to bind the two nuclei together, and regions of space in which the negative charge has the opposite effect. We refer to the regions as **binding** regions and **antibinding** regions, respectively. They are divided by "nonbinding" surfaces. A standing wave state for an electron, and for a particular pair of nuclei some distance R apart, may be classified as a bonding state or an antibonding state, depending on whether, when R is not too small, the electronic charge has the net effect of pulling the nuclei together or of keeping them apart. For sufficiently small R, electron distributions always become antibonding for two reasons: first, a pinched electron distribution has short wavelength, therefore high momentum, therefore high energy and therefore the capacity to escape; second, if two or more electrons are present the Exclusion Principle forces electronic charge out of the binding region of space. Thus we usually describe an electronic state of a molecule as bonding (with respect to any particular pair of nuclei, if there are more than two) if it distributes electron charge to make the long-range forces between the atoms attractive. Such a state has a stable equilibrium state at some value of R, called R_e, when attractive and repulsive forces, that is, bonding and antibonding forces, exactly balance. If the electronic charge distribution of a state makes the long-range forces between two nuclei repulsive, we say that state is antibonding with respect to those nuclei.

The Pauli Exclusion Principle sets a maximum of *two electrons* in any standing-wave state. This requirement is the physical basis for the long-established idea proposed by Gilbert N. Lewis that ordinary chemical bonds are formed by pairs of electrons. In most stable molecules each chemical bond is the result of just one pair of electrons occupying a bonding state. A chemical bond exists between two nuclei generally when the number of bonding pairs is greater than the net number of antibonding pairs. For most bonds that excess is just one pair. However in some examples, such as the nitrogen molecule, N_2, or the oxygen molecule, O_2, the number of bonding pairs is more than one greater than the number of antibonding pairs. In these cases, one speaks of **multiple bonds**. The N_2 molecule has *three* excess pairs of bonding electrons, and therefore is said to have a *triple* bond. Oxygen has only two more excess bonding pairs than antibonding pairs, and therefore is said to have a **double bond**. There are, of course, some exceptions to the electron-pair rule; molecules such as NO (nitric oxide) which contain odd numbers of electrons must necessarily have some state occupied by one single electron.

To break a typical single bond requires energy between about 2 and 5 electron-volts. In macro terms, it requires about 50 to 120 kilocalories or about 200,000 to 480,000 joules to break a mole of such bonds. Double bonds are considerably stronger, but typically not quite twice as strong. This is because the mutual repulsion of the electrons for each other tends to keep the electron density from concentrating in the most strongly binding regions of space. The triple bond is stronger still, but again, less than three times as strong as the single bond. A few typical bond energies of diatomic molecules are listed in Table 1.

TABLE 1

**Typical Bond Dissociation Energies of Diatomic Molecules
(at 298°K).**

Bond Type	kcal/mole	Joules
H–H	104	4.35×10^5
H–C	81	3.39×10^5
H–O	102	4.28×10^5
C=C	145	6.07×10^5
C≡O	257	10.76×10^5
O=O (as in O_2)	119	4.97×10^5
N≡N (as in N_2)	225	9.42×10^5
N=O	150	6.27×10^5
Na–Cl	98	4.10×10^5

In general, the stronger a chemical bond is, the more inert that bond is toward reaction. Oxygen, for example, is far more reactive than the very inert nitrogen, in part because of the difference in

124

their bond energies. The nitrogen molecule has one of the strongest bonds, and is one of the most inert molecules we know. The nitric oxide molecule, NO, falls about halfway between N_2 and O_2 in its bond strength, as the table shows.

AN ASIDE ON NITROGEN, OXYGEN AND NITRIC OXIDE

Nitrogen and oxygen are the principal constituents of air, which is about 80% N_2 and 20% O_2. Nitric oxide is formed when air is heated to 2000°K or more and cooled too fast for the NO to revert to N_2 and O_2. This happens in almost all kinds of combustion. Nitric oxide is a principal component of smog. Although the bond energy of NO is between those of N_2 and O_2, its bond energy is actually significantly less than half the average of the N_2 and O_2 bond energies:

$$150 < (225 + 119)/2 \ = 172 \text{ kcal/mole}$$

This means that internal energy is released when two molecules of NO form one molecule of N_2 and one molecule of O_2:

$$2NO \rightarrow N_2 + O_2 \ .$$

In fact 2 × (172 − 150) or 44 kcal of energy must be released whenever one mole of N_2 and one mole of O_2 are formed. (We shall return to this sort of arithmetic shortly, in the context of the First Law of Thermodynamics.) These energies are illustrated in Figure 22.

Ironically, despite the tendency of systems to run downhill to states of lower energy, two molecules of NO do not react readily to form $N_2 + O_2$; in fact, it appears to be exceedingly difficult to make this process take place. At moderately high temperatures, it is possible to bring this about with a catalyst, a substance that makes a reaction proceed more readily without itself participating in the net reaction. At quite high temperatures, such as those reached in

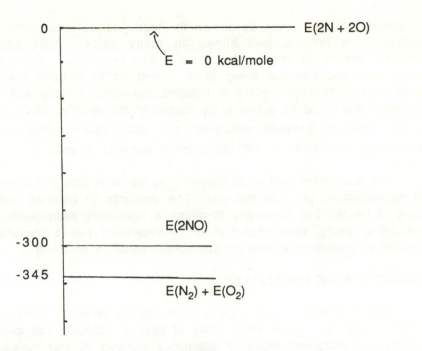

Figure 22. Energies of 2NO and $N_2 + O_2$.

hydrocarbon-air combustion chambers, the reverse reaction takes place fairly readily. Thus, in an automobile cylinder or in a furnace boiler, NO forms, and the system establishes an equilibrated mixture of $N_2 + O_2$ with NO. Then, when the hot gases pass out into the air and cool, they can no longer react at a measurable speed, so that the NO is, in effect, "fixed" into the system.

It is unfortunate that NO is a product of hot combustion reactions because it has a number of unpleasant consequences. Among them is its ability to react in air to form the brown gas NO_2. Both NO and NO_2 are irritating and damaging to plants; in addition, NO_2 can absorb sunlight to collect the energy required to drive the photochemical smog cycle. This is a complex series of reactions in which ozone is generated and hydrocarbons are oxidized to compounds that form irritating fogs called aerosols. Smog is easy to detect; its characteristic brownish haze, ozone odor and eye irritation are far too well known in many cities. One of the potential means to reduce the harmful and unpleasant aspects of smog, and perhaps the smog itself, would be to remove the NO formed in combustion before it reaches the open atmosphere. In principle, this could be achieved by means of the reaction $2NO \rightarrow N_2 + O_2$, but at present catalysts are only partly effective in stimulating this reaction with automotive exhaust gases.

An alternative method of control may be to reduce the amounts of hydrocarbons put into the air. The amounts of nitrogen oxides have to be reduced drastically to cause a significant decrease in the amount of smog, while reduction of hydrocarbons has a reasonable chance of producing a nearly-proportional reduction of smog.

ENERGY STORAGE IN MOLECULES

We have seen how energy is stored in the chemical bond itself. However this is not the entire story of how a molecule can contain energy; our excursion into heat capacities showed us that molecules should be able to vibrate and rotate. In fact, at normal temperatures, there are simply not enough quanta of high enough energy to allow us to see excitation of the electrons of a bond from their bonding state to some higher-energy standing wave state. To see these processes, we must go to extremely high temperatures or use special methods of excitation, particularly irradiation by ultraviolet light. Consequently, the electrons in an atom or molecule do not contribute to the normal heat capacity at room temperature because there are almost none of the very energetic quanta they require to become excited.

By contrast, molecules can absorb energy at typical room temperatures into their vibrational and rotational motion. Let us look at the potential energy curve for a molecule (Figure 23). Near the equilibrium internuclear distance R_e, this curve looks very much like a parabola, the potential energy curve of the harmonic oscillator. We therefore expect a molecule to vibrate much like a

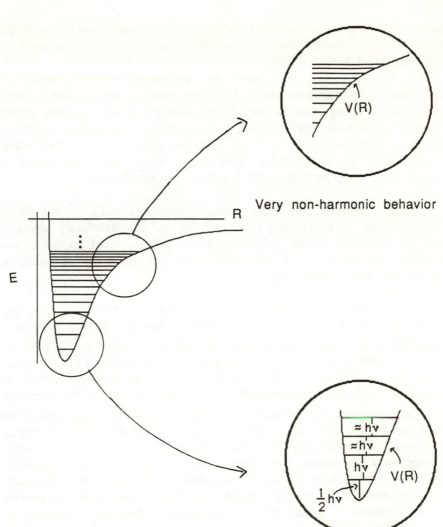

Very non-harmonic behavior

Nearly harmonic behavior

Figure 23. **Energy levels for molecular vibration.**

quantum-mechanical oscillator, with nearly equally-spaced energy levels. This is indeed the case, so long as the molecule remains in standing wave states of reasonably low energy, and therefore in the general vicinity of R_e. If we pump energy into the molecule so that a bond begins to oscillate wildly, then the system behaves as though the potential energy well is very broad, and the levels become closer together as the energy increases. At ordinary temperatures, or in the bottom of most typical bond potential wells, the standing-wave energy levels are separated by amounts of energy that correspond to quanta of infrared radiation ($10^{-2} - 10^{-3}$ cm wavelength).

The molecule also may contain quantized units of rotational energy, with level spacings that increase approximately linearly with the quantum number of the rotation. These level spacings are characteristically in the *microwave* region of the spectrum roughly 0.5 cm – 50 cm in wavelength.

The vibrational energy levels are responsible for the vibrational contribution of k_B per degree of freedom to the molecular heat capacity when the energy available in collisions is large compared with the difference in energy of one attainable vibrational level and the next. The rotational energy levels, similarly, contribute $k_B/2$ per rotational degree of freedom, when the average energy available in collisions is large compared with rotational level spacings. When the system is so cold that the average energy of molecular collisions is small compared with the vibrational energy level separation, the energy that can be exchanged between molecules is too little in almost all collisions to provide a quantum energetic enough to induce a quantum jump in the vibrational energy of a molecule. Under these conditions, vibrational degrees of freedom do not contribute significantly to the molecular heat capacity. By the same token, if the system is *very* cold, i.e. if the absolute temperature is very low, the energy available in translational motion is too low to excite even the quantized rotational motion of molecules. When this occurs, rotation ceases to contribute to the molecular heat capacity. Thus we can infer that a mole whose heat capacity at constant volume, C_v, is $^3/_2R + ^2/_2R + R$

or $^7/_2R$ at high temperatures, becomes $^5/_2R$ at somewhat lower temperatures and is only $^3/_2R$ at quite low temperatures, provided it does not condense first.

The fall-off in heat capacity at low temperatures was first interpreted for solids, by Albert Einstein. The essence of his interpretation, that the solid is a collection of oscillators with frequency ν, whose energy could only be contained in discrete units of energy $h\lambda$, was one cornerstone of the quantum theory of matter.

Problems

1. A diatomic molecule has a natural vibration frequency of 3×10^{13} sec^{-1}, and a rotational energy level separation between $J = 0$ and $J = 1$ of 5×10^9 sec^{-1}. Suppose it is cooled from a temperature of 1000°K, at which its heat capacity C_V is $^7/_2$ R. At about what temperature does vibration cease to contribute to the heat capacity of this species? At about what temperature does rotation cease to contribute?

2. In the photochemical cycle in which automobile exhaust products become smog, certain critical steps involve free oxygen atoms. These atoms come largely from nitrogen dioxide, NO_2, which has the structure given below. The process of oxygen atom production is this:

$$NO_2 + light \rightarrow NO + O \ .$$

One nitrogen-oxygen bond is thus broken when a quantum of sufficiently energetic light is absorbed. The process is much like ionization of a hydrogen atom by light, but the particle set free from NO_2 is an atom, not an electron. The energy required to dissociate, i.e. to break a nitrogen-oxygen bond has been measured. It is customarily reported in units of calories or kilocalories per *mole* of NO_2 molecules, or in electron-volts per molecule. This quantity is approximately 70,000 calories (70 kilocalories) per mole or 3 electron-volts per molecule. Its value is approximately 5×10^{-19} joules. (Is this per *mole* or per *molecule*?) What is the longest wavelength λ_{max} that light may have and be capable of dissociating NO_2 into NO and O? Is NO_2 dissociated primarily by visible, infrared or ultraviolet light? Ozone gas absorbs radiation very strongly in the spectral region from 200-350 nm (1 nm = 10^{-9} m) and weakly in the region 450–700nm. Is ozone an effective filter to prevent light-induced dissociation of NO_2?

13. ENERGY AND THE FIRST LAW OF THERMODYNAMICS

We have dealt now with energy in several ways: as a quantity associated with the mechanics of simple classical bodies and the forces between them, as a quantity associated with changes of state of a macroscopic collection of simple molecules, and, on a microscopic scale, as a quantity that exists in quantized units, whether contained in a radiation field or in some form of excitation of a molecular system. Now we turn to the general formulation of a notion that unifies all of our previous development, the First Law of Thermodynamics. The Laws of Thermodynamics are statements about the way we observe all the macroscopic world to behave. They are laws that derive their meaning from the very macroscopic nature of the systems they describe. They are, in that sense, laws about complex systems that arise partly from the mechanical laws that govern the microscopic matter comprising them, but even more from their own extreme complexity.

How can a concept or law be meaningful for a macroscopic system and not for a microscopic system? The answer is that to describe large, complex systems, we must use certain variables which are only meaningful in those systems, variables such as heat, temperature, entropy and even pressure. The First Law of Thermodynamics, in one form, is a statement that connects energy, work and heat. Both energy and work have meaning for large and small systems, and we have investigated both these properties in some detail, but heat – – what is heat?

We discussed heat in the context of heat capacities. There we stated, without delving into how it happens, that heat is energy distributed among all the degrees of freedom of all the molecules in a system, in that peculiar way which equipartitions the energy. When such a distribution occurs in an isolated, insulated system, the average energy in each degree of freedom is unchanging in time. This energy, so distributed, is as far as it can be from localization into a single form, which would maximize the system's capacity to do work.

Our best sources of work are those in which stored energy is highly organized and localized in a single degree of freedom. Heat is, in a sense, the least organized form that energy may have. Other kinds of energy, in such varied forms as a current of electrons, a flowing river, a sound wave or a laser beam, all consist of energy concentrated in a very few degrees of freedom. In an electric current, the energy is the kinetic energy of the electrons; in a river, the mass and therefore the gravitational potential energy and kinetic energy of flow are the storehouses of energy. Heat, by contrast, is energy shared among so many degrees of freedom that we cannot distinguish deviations or fluctuations away from equipartition among these degrees of freedom.

We shall later see how this characteristic of complex systems, the narrowness of range of fluctuations, is a direct consequence of the system's very large number of degrees of freedom. To anticipate our conclusion, we introduce our findings now. If a simple mechanical system with only a few degrees of freedom is left alone, either its energy remains in its original degrees of freedom or its energy flows from one degree of freedom to another. In either case, the energy distribution among degrees of freedom at any instant almost always deviates far from equipartition. We almost always observe large fluctuations away from the average distribution. As the number of degrees of freedom or the number of particles increases, these fluctuations in the energy distribution diminish until, for macroscopic collections of matter, they are immeasurably small under most circumstances. For large systems, we observe a constant, equipartitioned allocation of energy.

At this point it becomes meaningful to define **temperature**. Recall how we defined it previously, in terms of mean kinetic or translational energy of the particles. It measures the *intensity* or concentration of heat.

Temperature is only a useful concept if the energy in translation or in any other degree of freedom is essentially constant on the time scale of exchanges of individual quanta among degrees of freedom. We do not have much use for a thermometer that constantly fluctuates over a large range of temperatures. If the number of degrees of freedom of a system is large, this condition is fulfilled. If the number is small, as, for example, in the case of three or four coupled oscillators, then the energy exchanges among these degrees of freedom represent such large fluctuations that we have to be careful about how we use the idea of temperature in describing the system.

With this examination of the nature of heat and temperature, we can understand how laws stated in terms of these variables are specifically laws governing complex systems. This is in contrast to laws of mechanics, which are expressed in variables appropriate to the description of simple systems. In principle, we could express laws of thermodynamics in terms of mechanics, but the complexity required would be overwhelming; it would defeat the purpose of efficient expression of information that we expect of a scientific law. Rather, the bridge between thermodynamics and mechanics is made in a natural way through the application of *statistics* to the behavior of complex mechanical systems. We shall address this approach in Chapters 15, 16 and 17.

The First Law of Thermodynamics can be expressed in several ways; we shall consider three of its statements.

1. Energy is a conserved quantity in complex macroscopic systems, just as it is in simple mechanical systems.

2. All energy exchanged between a system and its surroundings must appear either as work, i.e. potential energy stored or used, or as heat. Let W be work done on a system, implying that a negative value of W means work is done by the system, and let Q be heat gained by the system. Then a change ΔE in the system's energy is

$$\Delta E = Q + W .$$

3. The energy of a system in a specific state is independent

of how that state was prepared. Hence any change of energy ΔE associated with a transition of a system between two specified states is a function only of the two states, and not of the path used to make the change. Whatever the input of heat Q involved in the particular pathway, the corresponding amount of work done by the system, $-W$, is fixed by the condition that: $-W = Q - \Delta E$.

The First Law has its microscopic interpretation in the same concepts that previously microscopic concepts to the notions of a macroscopic state and the equation of state. The state of a system of fixed composition corresponds to a condition in which the temperature and pressure or density are specified. These two restrictions fix the mean kinetic energy per particle and the mean interparticle distance, respectively. The mean interparticle distance, in turn, determines the amount of potential energy stored in interparticle interactions, as well as the momentum transferred to the walls each second. Any change in the energy stored as potential and kinetic energy of the particles must appear either as heat absorbed by these particles or work done on them, against their forces of interaction, just as energy is put into a spring by compressing it.

Problems

1. The mean fluctuations of energy per degree of freedom increase as $\sqrt{N}/N$ or as $1/\sqrt{N}$. Suppose a small system, containing 10 particles, has a mean fluctuation in energy per particle of 25%. What are the mean fluctuations in energy per particle for similar systems of 1000 particles? of 10,000 particles? of 10^{23} particles?

2. A falling raindrop has most of its energy in a single degree of freedom, the kinetic energy of its center of mass. When the raindrop strikes the earth, in what ways does its energy become divided among many degrees of freedom?

3. An ultraviolet lamp emits radiation of a single frequency, corresponding to that energy being in a very few degrees of freedom. The lamp is used to produce a "sun tan". How does the action of the lamp on a person involve conversion of the energy from a few degrees of freedom to many degrees of freedom?

14. ENERGY RELATIONS IN CHEMICAL PROCESSES: COMBUSTION

The First Law of Thermodynamics gives us a very convenient way to deal with conversion of chemical energy to heat, or the conversion of other forms of energy to stored chemical energy. This convenience comes from our ability to determine energy changes in very simple chemical steps and combine these changes to deal with quite complex substances.

BOND ENERGIES

As a first example, let us make use of the quantities known as **bond energies**. These are similar to the bond dissociation energies that we discussed in the context of the chemical bond, but are not strictly identical. The reason is that a *bond dissociation energy* is meant to be a precise, specific quantity, the energy required to break a particular bond in a particular sort of molecule. The exact energy to dissociate or break a bond depends somewhat on what its neighboring atoms are. In contrast, a *bond energy* is an approximation, an *average* figure for a variety of compounds. The two differ, but by amounts small enough that we can frequently neglect the differences. We do, however, have to distinguish single, double and triple bonds between the same kinds of atoms because these differ by considerable amounts. Table 2 gives some typical values of average bond energies for common bonds.

COMBUSTION

Here is how we use bond energies to estimate energy changes. Consider a real chemical reaction, the combustion of methane, CH_4:

<u>Reactant</u>   <u>Products</u>

$$CH_4 + 2O_2 \quad \rightarrow \quad CO_2 + 2H_2O \ . \qquad (71)$$

Experience tells us that this process releases heat; it is in fact the net chemical reaction that occurs in the ordinary clean burning of natural gas, which is almost all methane. The process may or may not be made to do work. It releases heat of course, but it does not do any net work when gas burns on a stove because the product gases return to the temperature and pressure of the surrounding atmosphere and, since the number of moles of product and reactant are the same, the product gases occupy the same volume as the reactants. How much energy is released as heat in the burning of a mole of methane?

Table 2

TYPICAL BOND ENERGIES

		kcal/mole	joules/mole
C=O	(average from two bonds of CO_2)	192	8.04×10^5
C=O	(as in acetone)	177	7.42×10^5
C–O	(as in CH_3OH, methanol)	90	3.77×10^5
C–H		99	4.15×10^5
C–C	(as in propane)	83	3.48×10^5
C=C	(as in ethylene)	146	6.12×10^5
C≡C	(as in acetylene)	200	8.38×10^5
C≡N	(as in hydrogen cyanide)	213	8.93×10^5
N–H		93	3.90×10^5
O–H	(as in CH_3OH)	111	4.65×10^5

To make use of the First Law in the methane combustion reaction, let us use the third form of the law, which states that the energy change is independent of the reaction path. We may therefore evaluate the energy released in the reaction by replacing **(71)** with a two-step process:

$$CH_4 + 2O_2 \rightarrow C + 4H + 4O \rightarrow CO_2 + 2H_2O \ . \qquad \textbf{(72)}$$

Here, the first step consists of breaking all the bonds of the reactants and thereby converting the entire system to atoms. The second step consists of letting these atoms form new bonds, the bonds of the product molecules. To accomplish the first step, we must supply energy equal to the sum of all the bond energies of the reactants:

Energy to be supplied = sum of all bond energies of reactants

or $\qquad E_{in} = \Sigma_{\text{reactant bonds}} (B.E.)_{\text{reactants}}$. $\qquad\qquad$ **(73)**

The second step releases an amount of energy equal to the sum of all the newly-formed bond energies of the products:

Energy released = sum of all bond energies of products

or $\qquad E_{out} = \Sigma_{\text{product bonds}} (B.E.)_{\text{products}}$. $\qquad\qquad$ **(74)**

The *net energy change* is the difference between these two energies. If we look at this energy change from the viewpoint of the system of burning methane and oxygen, then the energy *contained* in this system *decreases* in the process. The energy within this system rises by an amount $+E_{in}$ when we make atoms out of reactants and then drops by an amount $-E_{out}$ when we let the atoms make themselves into product molecules. Hence the change in energy of this system of molecules is

$$\Delta E \ = \ - E_{out} + E_{in}$$

$$= -\Sigma_{\text{product bonds}} (B.E.)_{\text{product}} + \Sigma_{\text{reactant bonds}} (B.E.)_{\text{reactant}}. \quad (75)$$

The Table of Bond Energies on page 137 can be used now to evaluate ΔE. The products have two C=O bonds and 4 O–H bonds, giving a product bond energy of

$$E_{out} = 2 \times 192 + 4 \times 111$$

$$= 384 + 444 = 828 \text{ kcal/mole of } CH_4 \quad (76)$$

The reactants have 4 C–H bonds and 2 O=O bonds, so

$$E_{in} = 4 \times 99 + 2 \times 119$$

$$= 396 + 238 = 634 \text{ kcal/mole of } CH_4 . \quad (77)$$

Hence the total energy change of the gas is

$$\Delta E = -828 + 634 = -194 \text{ kcal/mole of } CH_4 . \quad (78)$$

If the system does no work and has no work done on it, so that $W = 0$, then the heat *absorbed* by the system is –194 kcal/mole, or, because of the negative sign, the system *releases* 194 kcal to the outside world for every mole of CH_4 burned.

It is interesting to compare the energy released by combustion of methane with the energy released by combustion of larger hydrocarbons. Let us take octane, C_8H_{18}, as an example, and use our method of bond energies again. We replace

$$C_8H_{18} + 12\,^1/_2\,O_2 \rightarrow 8\,CO_2 + 9\,H_2O \quad (79)$$

(or, equivalently,

$$2\,C_8H_{18} + 25\,O_2 \rightarrow 16\,CO_2 + 18\,H_2O ,$$

if we wish to use only integer coefficients) with the two-step process

$$C_8H_{18} + 12\,{}^1\!/_2\,O_2 \rightarrow 8\,C + 18\,H + 25\,O \rightarrow 8\,CO_2 + 9\,H_2O \qquad (80)$$

Note that the use of non-integer coefficients does make physical sense if we regard the equations as referring to moles of reacting species, and not just to single molecules. The energies in (80) are

$$E_{out} = 16\,E(C=O) + 18\,E(O-H) = 3072 + 1998$$

$$= 5070 \text{ kcal/mole} , \qquad (81)$$

and $\quad E_{in} = 18\,E(C-H) + 7\,E(C-C) + 12\,{}^1\!/_2\,E(O=O)$

$$= 1782 + 581 + 1487.5$$

$$= 3850.0 \text{ kcal, so} \qquad (82)$$

$$\Delta E = -1219.5 \text{ kcal/mole of octane burned} \qquad (83)$$

Clearly, the energy released, per *molecule* or per *mole* is much greater when octane is burned than when methane is burned. But the released energy per molecule or per mole is often not the quantity of principal concern. This is particularly clear here because methane is so much smaller and lighter than octane. One is often interested in the energy released per unit *mass* of fuel, or for incompressible substances, i.e. solids or liquids, in the available energy per unit volume of fuel. Let us compare methane and octane on a weight basis. From CH_4, with a molecular weight of 16, we obtain

$$-\Delta E/\text{gram of methane} = 194 \text{ kcal/mole} \times {}^1\!/_{16} \text{ mole/gram}$$

$$= 12.1 \text{ kcal/gram of methane} . \qquad (84)$$

From octane, with a molecular weight of $8 \times 12 + 18 = 114$, we obtain

$$-\Delta E/\text{gram of octane} = 1219.5 \text{ kcal/mole} \times {}^1\!/_{114} \text{ mole/gram}$$

$$= 10.7 \text{ kcal/gram of octane} . \qquad (85)$$

Hence the extra energy released due to the relative weakness of the C–C bond is not enough to compensate for the added mass of the higher proportion of carbon in octane, and methane turns out to be the more energetic fuel on a weight basis.

Because of the ambiguities and experimental difficulties associated with defining bond energies, it is often useful to use a conceptual stepwise process slightly different from our bond energy approach in order to evaluate energy changes in chemical reactions. Instead of an intermediate state consisting of isolated atoms, let us use an intermediate state consisting of the pure elements in their normal state at room temperature and atmospheric pressure. This way, we replace the conceptually simple but experimentally inconvenient atomization process with the much more realistic and practical process of referring all compounds to the elements of which they are composed. The standard **energy of formation** of a compound is the energy change of the materials, which comprise our compound when they form that compound from its elements as they ordinarily occur.

Because we work more often at constant pressure than at constant volume, it is more convenient to deal with enthalpies than with energies. The difference is that the enthalpy is defined so that its change omits the energy released as work of expansion against the constant external pressure. Hence a change in the enthalpy tells us the amount of energy released as heat in a constant-pressure process. The enthalpy change in a reaction in which the reactants start and the products finish at room temperature and atmospheric pressure is called the **standard heat of reaction**, and if the reactants are pure elements, then these energies are called **standard heats of formation**, which we shall symbolize as ΔH_f.

ENTHALPY

The enthalpy H is defined as

$$H = E + PV \qquad (86)$$

Here, P is the pressure exerted *by* the system or the internal pressure, and V is its volume. Since E, P and V are functions of state and not of path, H is also a function of state, just as the energy is. Thus a change ΔH in the enthalpy, between two states, can be written

$$\Delta H = \Delta E + \Delta(PV) \qquad (87)$$

and if the internal pressure P is constant, then the enthalpy change is

$$\Delta H = \Delta E + P\Delta V . \qquad (88)$$

But the First Law lets us replace ΔE with $Q + W$, so

$$\Delta H = Q + W + P\Delta V .$$

Now recall that the mechanical work done by a system against an *external* pressure p is $p\Delta V$ (See Eq. (17)). Therefore

$$= Q - p\Delta V + P\Delta V . \qquad (89)$$

Then *if* the external pressure p and the internal pressure P are essentially equal, the two terms involving ΔV are equal and appear with opposite signs:

$$\text{if} \qquad p \approx P , \qquad (90)$$

$$\text{then} \qquad p\Delta V = P\Delta V \qquad (91)$$

and the enthalpy change of the system is the same as the heat absorbed by the system and

$$\Delta H = Q , \text{ if condition (90) is fulfilled} . \qquad (92)$$

The condition that $p \approx P$ is the same as the condition that the system never be more than an infinitesimal way from being in equilibrium with its surroundings, which we recognized as defining a *reversible* process. This was the condition for the *most efficient possible extraction of energy as work.* But if the system does the maximum amount of work in going between two particular states in some specified way, i.e. if −W is as large as possible within the specifications we place, then the First Law requires that Q be as large as possible because ΔE, which fixes the sum Q + W, depends only on the end-points of the path. Hence ΔH is really the maximum energy that the system can exchange as heat in a reaction at constant pressure.

RETURN TO COMBUSTION

Now let us go on to use the standard heats of formation to reevaluate the heat of combustion of methane. We replace the one-step reaction **(63)** with a two-step process, but instead of the atomization step of **(64)**, we now use this sequence:

$$CH_4 + 2\,O_2 \rightarrow C + 2\,H_2 + 2\,O_2 \rightarrow CO_2 + 2\,H_2O \ .$$

The standard heat of formation of O_2 is zero, as one would expect since it is an elemental molecule (composed of a pure element), and represents the normal form of the element at room temperature and atmospheric pressure. The heats of formation of CO_2, H_2O and CH_4 are quantities derived from precise and unambiguous experiments; we find that

$$\Delta H_f(CO_2) \ = \ - \ 94 \ \text{kcal/mole} \ ,$$

$$\Delta H_f(H_2O) \ = \ - \ 57.8 \ \text{kcal/mole} \ ,$$

and $$\Delta H_f(CH_4) \ = \ - \ 17.9 \ \text{kcal/mole} \ .$$

This reaction has no change in the number of moles of gas, so Δ(PV) is zero. If the system starts and finishes at room temperature and pressure, the total change in enthalpy and, in this case, in energy is

$$\Sigma_{\text{products}} \, \Delta H_f - \Sigma_{\text{reactants}} \, \Delta H_f \; =$$

$$[(-94) + 2 \, (-57.8)]_{\text{products}} - (-17.9)_{\text{reactants}} \; =$$

$$191.7 \text{ kcal/mole of } CH_4 \text{ burned} \; . \qquad\qquad \textbf{(93)}$$

This value is near but not identical to the value we obtained using the bond energy method. The difference between the −191.7 kcal/mole calculated here and the −194 calculated previously is a reflection of the inaccuracies introduced into the bond energy method because of the way we force a model beyond its true range of validity. "The" bond energy is only a concept valued as an approximation or an average and therefore is useful only in that sense. Often in science we deliberately use an approximate model because it is far more convenient than the next more accurate procedure. In many instances, the results of the crude model are accurate enough to answer the question at hand. Were we to demand 5% accuracy in the heat of reaction **(71)**, the bond energy method would suffice; were we to demand 0.5% accuracy, we would have to use heats of formation. Actually these are usually known to one or two more significant figures than we have quoted here.

We conclude this chapter by making estimates of the maximum temperatures that can be attained in a methane-oxygen flame and in a methane-air flame. We know that about 192 kcal are released when a mole of methane burns. When it burns, and this heat is released, the product gases, CO_2 and H_2O, may absorb that heat. Alternatively, some or all of the heat may be lost to the surroundings. Now, however, we want to examine the extreme situation, in which the product gases reach their maximum temperature. Let us also assume that the product gases stay at a pressure of 1 atmosphere, corresponding to the most usual conditions of flames.

How do we go about coping with this rather formidable looking question? In fact, it is quite straightforward, as we can see by breaking the problem into components: what do we know and what are we asked? We *know* how much heat we must distribute into the products of burning each mole of methane. Moreover we know precisely what those products are: one mole of CO_2 and two moles of H_2O, according to Eq. **(71)**. If our reactant mix consists of only oxygen and methane, then a mix of exactly two moles of oxygen for every mole of methane will provide the greatest amount of available heat energy, per molecule in the final system. Any other mix will contain some unreacted starting material in the products, which will absorb some of the generated heat. In the exact 2:1 mix, all the methane gets burned and there is no excess oxygen left over to blot up heat in the product system. This precise mixture is called a *stoichiometric* mixture. If there is extra methane the mixture is said to be *rich* or *fuel-rich*; if there is an excess of oxygen (whether it appears as pure oxygen or in air), the mixture is said to be *lean*. Suppose we are dealing with a stoichiometric mixture. In this case we know that the 192 kcal of heat energy must be distributed to precisely one mole of CO_2 and two moles of H_2O, so long as we use $CH_4 + O_2$ to make our flame.

How can we infer a temperature if we know what the species are and how much heat is available? Let us recall our discussion in Chapter 6 on heat capacities of gases. The CO_2 molecule is linear, and H_2O is nonlinear. Hence CO_2 must have $^3/_2 \mathbb{R}$/mole-deg of translational heat capacity, $2 \times ^1/_2 \mathbb{R}$/mole-deg of rotational heat capacity, and $\mathbb{R}$ for each of its $3n - 5$ degrees of vibrational freedom, where n, the number of atoms, is 3. Therefore CO_2 has a heat capacity at constant volume, C_V, of $^3/_2 \mathbb{R} + \mathbb{R} + 4\mathbb{R} = ^{13}/_2 \mathbb{R}$ cal/mole-deg. The heat capacity C_p, for heat changes at constant pressure, is larger by an amount $\mathbb{R}$, and $\mathbb{R}$ has the value 2 cal/mole-deg, so

$$C_p(CO_2) = {}^{15}/_2 \mathbb{R} = 15 \text{ cal/mole-deg} . \qquad \textbf{(94)}$$

Similarly, the nonlinear molecule H_2O has $^3/_2 \mathbb{R}$ cal/mole-deg of translational heat capacity, $^3/_2 \mathbb{R}$ cal/mole-deg of rotational heat capacity (with three axes of rotation, not two), and $(3n - 6)\mathbb{R}$ or $3\mathbb{R}$

cal/mole-deg of vibrational heat capacity. Therefore gaseous (vapor) H_2O has a constant-volume heat capacity C_V of $^{12}/_2 \mathbb{R}$ or 12 cal/mole-deg, and a constant-pressure heat capacity

$$C_p = 14 \text{ cal/mole-deg} . \tag{95}$$

Therefore the *total* heat capacity of the product gases from burning one mole of CH_4 in two moles of O_2 is

$$15 \text{ (from } CO_2\text{)} + 2 \times 14 \text{ (from } H_2O\text{)} =$$

$$43 \text{ cal/mole of } CH_4\text{-deg} . \tag{96}$$

The total rise in temperature, ΔT, must be related to the heat Q by the relation

$$\text{total heat Q} = \text{(total heat capacity)} \times \Delta T$$

or $\quad \Delta T = \text{total heat/total heat capacity} . \tag{97}$

Therefore $\quad \Delta T = \dfrac{192,000 \text{ calories/mole of } CH_4}{43 \text{ calories/mole of } CH_4\text{-deg}}$

$$= 4465° \tag{98}$$

and the final temperature is

$$T = \Delta T + \text{room temperature } (300°K)$$

$$= 4765°K , \tag{99}$$

which is easily hot enough to melt virtually any material. (Actually the losses by radiation reduce the temperature significantly.)

If we use air instead of pure oxygen, the temperature must be a good deal lower, simply because air is 80% nitrogen and, apart from the formation of NO that we discussed previously, the N_2 acts only as an inert energy absorber. How much heat capacity does the nitrogen contribute? For every mole of O_2 in air, there are *four* moles of N_2. Each mole of N_2 has $^3/_2 \, \mathbb{R}$ of translational heat capacity, $^2/_2 \, \mathbb{R}$ of rotational heat capacity, $\mathbb{R}$ of vibrational heat capacity, and another $\mathbb{R}$ for the expansion associated with a process at constant pressure, giving us $^9/_2 \, \mathbb{R}$ or 9 cal/mole-deg. This adds 4 × 9 or 36 calories to the heat capacity of the gases from combustion of one mole of CH_4, so that, with air instead of oxygen,

$$\Delta T = 192,000/79$$

$$= 2430° \ ,$$

and
$$T = 2430 + 300$$

$$= 2730°K \ . \tag{100}$$

Thus, even with air instead of pure oxygen, the temperatures are very hot indeed. The losses due to radiation are considerable in an ordinary burner flame, but the combustion products in an automobile cylinder do typically reach temperatures in the vicinity of 2500 – 2700°.

Problems

1. The water molecule, H_2O, is triangular. Carbon dioxide, CO_2, is linear. How many calories are required to heat one mole of water vapor from 300°K to 1000°K? How many calories are required to heat one mole of CO_2 from 300°K to 1000°K? (Assume the heating is done at constant pressure.)

2. The bond in the H_2 molecule has an energy of 4.5 electron volts, equivalent to 104 kilocalories of energy, per mole of H_2. The bond in the O_2 molecule has an energy of 5.1 electron volts, corresponding to an energy of 119 kilocalories per mole. The average energy of the O–H bonds of H_2O is 4.8 electron volts or 103.5 kilocalories/mole. How much net energy is released from the chemical bonds when two moles of H_2 react with one mole of O_2 to form two moles of water, according to the reaction

$$2\,H_2 + O_2 \rightarrow 2\,H_2O \ ?$$

Suppose all the energy of the reaction is used to heat the product (water vapor) at constant pressure. What temperature does the water vapor reach?

3. Using the bond energies in Tables 1 and 2, and the principle of conservation of energy (the First Law of Thermodynamics), compute the energy released by each of the following combustion reactions, when <u>one mole</u> of fuel (methane, propane or ethyl alcohol) is burned. What is the mass of one mole of each fuel? The densities of the fuels, stored as liquids, are as follows:

methane, CH_4	0.41 gm/cm^3
propane, C_3H_8	0.58 gm/cm^3
ethyl alcohol, C_2H_6O	0.79 gm/cm^3

List these three fuels in order of decreasing energy of combustion per gram, and then in order of decreasing energy of combustion per cm^3. On the basis of methane and propane, generalize about the way fuel value – both as per unit mass and as per unit volume – changes as the ratio

number of hydrogen atoms/number of carbon atoms

decreases. How does methanol (methyl alcohol) compare with methane or propane in terms of the energy it can supply, per unit mass and per unit volume?

$$\overset{\displaystyle H}{\underset{\displaystyle H}{H-C-H}} \quad + \quad 2O_2 \quad \rightarrow \quad CO_2 \quad + \quad 2H_2O$$

methane   oxygen   carbon dioxide water

$$\overset{\displaystyle H \quad H \quad H}{\underset{\displaystyle H \quad H \quad H}{H-C-C-C-H}} \quad + \quad 5O_2 \quad \rightarrow \quad 3CO_2 \quad + \quad 4H_2O$$

propane

$$\overset{\displaystyle H \quad H}{\underset{\displaystyle H \quad H}{H-C-C-O-H}} \quad + \quad 3O_2 \quad \rightarrow \quad 2CO_2 \quad + \quad 3H_2O$$

ethyl alcohol

4. Explain why the generally accepted form for tabulating the combustion properties of fuels is in terms of *enthalpies* of combustion rather than *energies* of combustion, when either would be logically satisfactory.

5. Consider the combustion of methane, as stated in Problem 3. If the water were to remain as vapor, and the volumes of both reactants and products were measured at 0°C and 1 atm pressure, what would the change of volume be, per mole of methane, as a result of combustion? If the water condenses to

a volume small enough to be neglected, what is the change in volume, per mole of methane, due to combustion? (Again, take reactants and products at 0°C and 1 atm pressure.) In the case of water condensing, what is the change in pV? What is the enthalpy of combustion, $\Delta E + \Delta(pV)$, per mole, for methane?

15. DISTRIBUTIONS

The discussion thus far has dealt, on one hand, with states of equilibrium and the near equilibrium states associated with reversible processes. On the other hand, when we examined systems by looking at properties of all the component particles, as with the kinetic theory of gases, we represented them by making arbitrary (but convenient) approximations – – like giving equal speeds to all particles – – that could not possibly correspond to states in equilibrium.   In deriving the equation of state for ideal gases, we assumed not only that all the molecules travel with the same speed but that $1/_6$ of the molecules move toward each wall at any instant. In reality such an assignment of velocities could at most persist for the time required for a few collisions. One would soon find some of the molecules moving slower than the average, some moving faster than the average and almost none moving at precisely the average speed.   Moreover the real system would almost instantly set molecules moving randomly in all directions, instead of perpendicular to the walls.

The assignment of velocities or kinetic energies among all the moving particles of a gas system is called the *velocity* or *energy distribution* of the system. Likewise the assignment of the number of energy quanta to each member of a set of identical harmonic oscillators is the quantum or energy distribution for that system of harmonic oscillators. A distribution tells how the intensity of some property is spread among members of a set.   For example, a population graph, showing how many members of the population have a particular amount of a property – – energy, money, or age – – is an energy, money, or age distribution. We are already familiar with some such curves: the age distribution of a population is the number of people of a given age, as a function of age; the income distribution is the number of people having a given income, as a function of income. We have already seen one such distribution for a physical system, the distribution of the energy stored in oscillators of different frequencies in a radiation field, as shown in Figure 16.

Real systems have the remarkable property that, when they are in equilibrium, they exhibit a very specific distribution among the available quantum states that depends only on the nature of the quantum states of the system and on the variables required to specify the macroscopic state of the system. These, we recall, are the variables that enter into the equation of state. The quantum levels are determined by the nature of the species making up the system and on the boundary conditions. Natural macroscopic state variables of a system of a fixed amount of a single pure substance are the temperature and volume, or any other two variables that can be inferred from these two via the equation of state. (If the system consists of more than one substance, then one more variable, such as percent composition, must be added for each new substance after the first.) One of the main objectives in this section is understanding how an isolated system prepared according to some arbitrary microscopic distribution will always approach the equilibrium distribution characteristic of that system, with its particular volume and total energy, or temperature, or pressure.

What are the characteristics of the distribution associated with an equilibrium state? Consider first some of the simple systems we have examined, such as the simple harmonic oscillator and the particle in a one-dimensional box. The allowed energy levels of a harmonic oscillator are equally spaced, at intervals $h\nu$, proportional to the oscillation frequency ν characteristic of the oscillator. It makes no sense to speak of a distribution for one simple physical system; we must imagine a collection of many copies of that system in order to discuss distributions. The name given to such a collection is an **ensemble**. In an ensemble consisting of a large number N of identical oscillators, how is the population of the system distributed with respect to the number of quanta in each oscillator? How many have no quanta, how many have one, how many have two, and so forth? The answer, in a qualitative sense, is shown in the curve of Figure 24. In more general, quantitative terms, the number of oscillators having a number j of quanta is $n(j)$, and is given by the decreasing exponential distribution

$$n(j) = n(0)e^{-\text{const} \times j} \qquad (101)$$

where the constant is $h\nu/k_BT$, so that $jh\nu/k_BT$, the exponent, is proportional to the total energy separating state j from state zero.

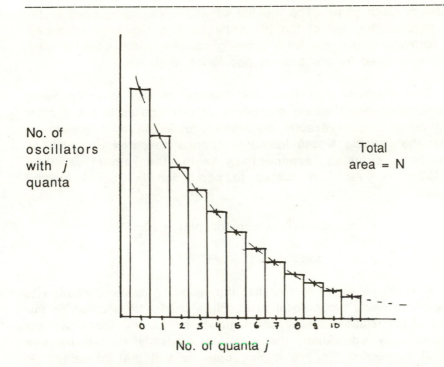

No. of
oscillators
with j
quanta

Total
area = N

No. of quanta j

Figure 24. **The distribution function for a system of N harmonic oscillators.**

The origin of Eq. **(101)** is in its essence the same as the origin of the human population growth with time or the radioactive decay over time of the amount of radioactivity in any given sample. In all these cases, the *fractional* change in a population variable is proportional to the absolute change in another variable. Human population itself – – number of people – – exhibits a *fractional change* or *percent change* that increases with time, and in a manner approximately proportional to the change in time itself; that is, the *percent* change in population during each year is apparently nearly constant over periods of many years. By contrast, radioactive species *decay*, atom by atom, with a fixed fraction decaying each second or each year. The half-life of any decaying species is the time required for half of the presently existing members to decay. The corresponding figure for a growing species is the doubling time, the time required for the present population to double.

That relation of a fractional change in one quantity being proportional to the *absolute* change in another applies to many other distributions. It is exactly equivalent to exponential growth or decay: the variable whose fractional change concerns us, such as population, changes exponentially with the other variable. Population in year 7 is related to population in year 1 by the equation

$$\text{population}(7) = \text{population}(1) \times 10^{\text{constant} \times (7-1)}$$

or

$$\text{population}(1) \times e^{\text{constant} \times (7-1)}.$$

It would be unfair to say that the relation (fractional change in one variable being proportional to absolute change in another) is the *cause* of the relation of exponential growth or decay. These two are mathematically equivalent; they are not causally related by any physical connection that we might adjust to test that causality. If the fractional change/absolute change proportionality describes a situation, then that situation must exhibit exponential growth or decay and conversely, any system showing exponential change must have fractional change in one variable proportional to absolute change in the other.

The exponential behavior occurs with other variables than time; it appears with variables such as height, energy and quantum number. The density of any planet's atmosphere falls off exponentially with altitude. The height at which the atmospheric density is half the density at the planet's surface is a good characteristic measure of the rate of decrease of density with altitude, and thus of the depth of that atmosphere. Here, density plays a role analogous to that of the number of radioactive atoms in our earlier example; density specifies the number of molecules per unit volume at every height. The characteristic altitude, for earth, where the density is half that at the surface, is about 5.5 km. Another example of an exponential fall-off is the distribution of a fixed number of quanta among the members of an ensemble of identical oscillators. This distribution also has a characteristic measure, the number of quanta of that state for which the population is half the population in the state of lowest energy.

Other distributions characterized by a single property also display exponential drop-off when the number of species in each state is plotted against the amount of that property. Consider the distribution associated with the energy of the state. The general form of the population distribution among quantum states, specifically of the number of particles N(E) with energy E, in a one-dimensional system, is

$$N(E) = N(0)e^{-E/k_BT} . \qquad (102)$$

We call N(0) the number in the lowest quantum state. (Restricting ourselves to a one-dimensional system eliminates the problem of counting several states of the same energy.) The harmonic oscillator distribution has a particularly simple exponential behavior as a function of the quantum number or the quantum state index j, as well as of the energy E_j, because the energy E_j of the jth state is directly proportional to the quantum number j.

The dependence on energy of the fractional change in each slice of the distribution can be put into the precise terms of an equation, as well as into an English sentence. If $\Delta N(E)$ is the change in the population distribution from energy E to a slightly higher energy E + ΔE, and N(E) is the population or number of species with energy E, then the change ΔN reflects a decrease in population (hence the negative sign) with an increase in E, according to

$$\Delta N(E)/N(E) = -(\text{positive constant}) \times \Delta E , \qquad (103)$$

where the positive constant is the quantity $1/k_BT$, the inverse of ($^2/_3$ of) the mean kinetic energy, per particle. This equation says that the fractional decrease in population is proportional to the absolute increase in internal energy, measured in units of k_BT, mean energy available to each degree of freedom of the system, when it is at a temperature T. The higher the temperaure T, the greater is the energy $k_BT/2$ available for each degree of freedom, and the smaller is the ratio $\Delta E/k_BT$, for any chosen ΔE. But the smaller $\Delta E/k_BT$ is, the smaller is the *fractional decrease* $\Delta N(E)/N(E)$ with the energy change ΔE, and the flatter is the population distribution. Thus high temperatures imply broad distributions and permit the species to be spread over a wide range of individual energy states. Low temperatures lead to narrow distributions peaking sharply at the lowest energy state. The forms of low-temperature and

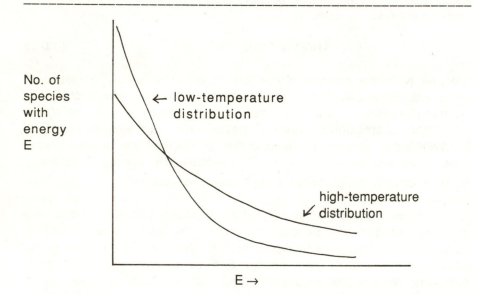

Figure 25. Low-temperature and high temperature distributions for a one-dimensional system of N particles. In both cases, the area under the curves is N.

high-temperature distributions for a one-dimensional system are shown in Figure 25. If an ensemble is cold, there simply are not enough quanta for any system to have a good opportunity to hold very many of them.

Thus far we have dealt only with distributions in one dimension, in which the number of systems in the state j, corresponding to energy E_j, is proportional to e^{-E_j/k_BT}. We can interpret this expression slightly differently to mean that the probability of finding a species selected at random in the state j, corresponding to energy E_j, is proportional to e^{-E_j/k_BT}.

Now let us consider atoms or molecules or oscillators that exist in the real world of three dimensions. It would also be a realistic problem to consider the case of species on a surface, which have only two dimensions in which to move. If the energy E_o of the lowest state is taken as zero, then $\Delta E_j = E_j - E_o = E_j$. With this choice, the probability of finding a molecule with energy E_j must surely be proportional to e^{-E_j/k_BT} in this three-dimensional case, if it is true in one dimension – – and in fact it is. However this is only part of the story.

In dealing with one-dimensional systems, we always speak of *the* state j with energy E_j implying that there is only one such state. However in two or more dimensions, where a system may have two or more equivalent degrees of freedom there may be more than one quantum state having the energy E_j. We already introduced the name for this; the number of quantum states of a given energy is the *degeneracy* of that energy level. The reason for this multiplicity or degeneracy always lies in the inherent symmetry of space or of space-time, and hence the equivalence of degrees of freedom. Let us symbolize by g_j the degeneracy of the level at the energy E_j. Then there are g_j distinguishable quantum states, characterizable by their own quantum numbers, all corresponding to species having energy E_j – – or, alternatively, there are g_j different and independent ways to prepare a species with the energy E_j.

    If there are g_j ways to prepare a molecule with energy E_j, then we would expect each of these ways to carry a probability corresponding to the probability we found in the one-dimensional case. The *total* probability of finding a selected molecule in the jth energy level should be the sum of the probabilities of finding each independent state. But because, among those j states, the individual probabilities are identical and there are g_j of them, the total probability of finding any randomly-selected molecule with energy E_j is proportional to the product of the degeneracy and the exponential:

$$P(E_j) \propto g_j e^{-E_j/k_B T} . \tag{104}$$

The constant of proportionality needed to make this into an equation is the number that assures all the $P(E_j)$'s add to unity, so that they behave like probabilities. This is the inverse of the sum of all the quantities $P(E_j)$. We call this the sum over states

$$Z = \Sigma_{\text{all states}} P(E_j),$$

which is also known as the *partition function*. The probability of finding E_j is then

$$P(E_j) = Z^{-1} g_j e^{-E_j/k_B T} . \tag{105}$$

    The three-dimensional case differs from the one-dimensional case simply because of the factor g_j. We recall that g_j generally increases with increasing energy. For example the degeneracies of the levels E_j of the hydrogen atom are given by $g_j = j^2$, and the degeneracy of the jth energy level of a rigid rotator is $g_j = 2j + 1$. The effect of g_j's increasing with increasing energy is to provide more ways to achieve states of higher energy, thereby increasing the probability for the higher energies and increasing probable populations. In fact the populations of energy levels in three-dimensional systems *do* increase with energy, up to energies E_j of about $k_B T$. The populations of states whose energies are greater than $k_B T$ decrease with increasing energy. This is shown graphically in Figure 26. The mathematical basis of the shape is dominance by the increasing function g_j for low energy and dominance by the decreasing exponential function at high energy. The physical basis of the shape is in the increasing number of ways to attain E_j, at low E_j, and the decreasing availability of energy to achieve E_j when E_j is large.

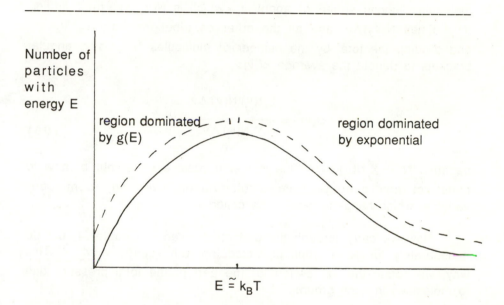

Number of particles with energy E

region dominated by $g(E)$

region dominated by exponential

$E \cong k_B T$

Energy $\rightarrow$

Figure 26. **Population distribution for a system of N particles in a three-dimensional space. The area under the curve is N.**

As an example of the application of distributions, let us reexamine the derivation of the ideal gas equation, which we carried out in Chapter 4. As before, we assume that one-sixth of all the molecules move toward each of the six walls of the rectangular parallepiped container, but now we suppose that the speeds v of the molecules are distributed over all possible values according to a distribution, in this case a *speed* distribution. If this distribution is $N(v)$, then $N(v_1)\Delta v$ is the number of molecules in the box having speed in the range $v_1 \pm \Delta v/2$, i.e. with speeds lying in a band Δv wide around the speed v_1. We compute the average value of any function of v, say f(v), by multiplying the value f(v_1) by the number of molecules having speed of about v_1, which is $N(v_1)\Delta v$, then adding f(v_2) times $N(v_2)\Delta v$, and all the other contributions from v_3, v_4 ... and dividing the total by the number of molecules N. Using angular brackets to denote the average $<f(v)>$,

$$<f(v)> = \frac{\sum_j f(v_j) N(v_j) \Delta v}{\sum_j N(v_j) \Delta v}$$

(106)

because $N = \sum_j N(v_j) \Delta v$. This is just a precise statement of how to construct any average from a distribution over any continuous variable, which happens here to be called v.

Now we carry through the derivation given in Chapter 4, but do it separately for each group of molecules with speeds v_1, v_2 .. This way, we obtain the expression for the part of the total pressure due to molecules in each group:

$$P_j = (^1/_3)\frac{N(v_j)}{V} \Delta v\, mv_j^2 .$$

(107)

The total pressure is obtained by adding the contributions of all the groups:

$$P = \sum_j P_j = (^1/_3)\frac{m}{V}\sum_j v_j^2 N(v_j) \Delta v .$$

(108)

The summation in this expression is precisely the total number of molecules N multiplied by $<v^2>$ the average value of the function

$f(v) = v^2$, so

$$P = \frac{(1/3)mN<v^2>}{V}$$

(109)

just like Eq. **(20)**, except that $<v^2>$ replaces v^2. This discussion shows how we can replace a crude, order-of-magnitude argument, such as the one used to establish Eq. **(20)**, with a far more accurate, physically plausible but more complicated argument that puts the same conclusion on a precise, logically sound basis. Successions of arguments like this occur frequently in science, as scientists work from rough, intuitive ideas to more refined theories reliable enough to make quantitative predictions.

Problems

1. Using a hand calculator or computer, calculate and graph the probability distributions of population at thermal equilibrium among the energy levels of a one-dimensional harmonic oscillator $N(v) = e^{-n(hv/k_BT)}/\Sigma_{j=0\to\infty} e^{-j(hv/k_BT)}$ for three cases:

   a) $hv/k_BT = 2$, the vibrational spacing hv is twice the thermal energy k_BT;

   b) $hv/k_BT = 1$, the vibrational spacing and the thermal energy are equal;

   c) $hv/k_BT = 0.2$, the vibrational spacing is one-fifth the thermal energy.

   In evaluating the denominator, the sum of course cannot be carried to infinity. It suffices to carry it only to a value of j high enough to make the first three significant figures constant. An alternative way to evaluate the sum is to use the fact that

   $$\Sigma_{j=0\to\infty} e^{-j(hv/k_BT)} = 1 + e^{-hv/k_BT} + (e^{-hv/k_BT})^2 + (e^{-hv/k_BT})^3 + \ldots$$

   $$= 1/(1 - e^{-hv/k_BT}) \ .$$

   This is simple and convenient; if you use it, show that it is true; that is, show that $1 + x + x^2 + x^3 + \ldots = 1/(1 - x)$ where, in our particular case, we have set $x = e^{-hv/k_BT}$.

2. Carry out the same calculation as in Problem 1 above, but for a two-dimensional harmonic oscillator. The only differences here from Problem 1 are that the degeneracy of each level g_j is exactly equal to $j + 1$, and therefore the number of oscillators expected to be found in level n is

$$N(n) = g_n e^{-n(h\nu/k_BT)}/\Sigma_{j=0\to\infty} g_j e^{-j(h\nu/k_BT)}$$

$$= (n+1)e^{-n(h\nu/k_BT)}\Big/\Sigma_{j=0\to\infty}(j+1)e^{-j(h\nu/kT)} \; .$$

Do the calculation and the graph for all three values of $h\nu/k_BT$. You should compute the approximate value of the denominator numerically, for all three values.

16. MICROSTATES, MACROSTATES AND ZERMELO'S PARADOX

Now we are prepared to ask why real systems in thermal equilibrium always choose the specific kind of distribution that they do. Why can we prepare a system with any arbitrary distribution and watch it evolve in time toward its equilibrium like that in Figure 26? (Such a distribution is often called "Maxwellian", after James Clark Maxwell, 1831–79). Why do we not see systems evolving into all the possible distributions, but rather, see only the equilibrium distributions?

Another way of phrasing the same question is in the form known as Zermelo's Paradox. As a syllogism, it can be put this way.

a) Nature presumably follows the Laws of Mechanics.

b) The Laws of Mechanics are time-reversible, in the sense that any mechanical event as it actually occurs is just as valid as the event that would have occurred if all the velocities of all the particles were reversed. Hence any mechanical event and its time-reversed image should be equally probable events in nature and one should be no more "natural" than the other. In everyday language, a movie representing a simple, mechanical process is just as valid a representation as the same movie run backward.

c) Therefore we should be unable to distinguish a direction for time, simply by observing natural events.

However, a movie of a plate dropping and breaking, simple as that event seems, is plausible, but the same movie run backward is not. Everyday experience obviously conflicts with the inference of this seemingly logical argument. The contradiction of experience with the conclusion of our syllogism is unavoidable. How is it that Nature can be governed by time-symmetrical mechanics, yet exhibit such a clearly marked directionality for time?

To probe this dilemma, to interpret the nature of equilibrium and nonequilibrium distributions, and to understand a little about evolution in time, we must examine the nature of the "state". Let us start by once again distinguishing between a macrostate, which is a condition of a macroscopic system specified by the variables contained in the equation of state, and a microstate. A **microstate** is a condition in which we have specified some or all of the variables associated with the individual particles, the atoms and molecules, of the system. In the extreme, we would specify all the quantum numbers of all the particles, and thereby specify a single pure microstate. The microstate is *far* more specifically determined than a corresponding macrostate. Knowledge of a microstate consists of as much information about the system as mechanics, classical or quantum, allows us, consistent with the way we have prepared the system. Normally, we have little or no interest in the microstates of bulk matter because there are so many of them, and because each one is specified by far more information that we could ever handle. Instead, and remarkably, we developed the variables of the macroscopic state – – temperature, pressure, volume, and the more abstract quantities such as energy and enthalpy – – which serve superbly as summary variables.

We shall now look at some simple, small systems to see how the number of microstates of a system is related to possible macrostates, as a function of size, i.e. the number of particles, comprising the system. To illustrate, let our system be composed of dice. As in the classic game of Craps, the total number showing on the top faces of the dice defines the macrostate. Each particular combination of numbers on the individual dice defines a microstate. If our system is only one die, then all the microstates are 1, 2, 3, 4, 5 and 6, and these are the same as the macrostates because there is only one way to achieve each number.

Suppose the system consists of two dice. Then the macrostates are defined by the values 2, 3, 4, . . . 10, 11, 12. Figure 27 shows how each macrostate can be made in terms of microstates, and shows also the number of microstates corresponding to each macrostate.

166

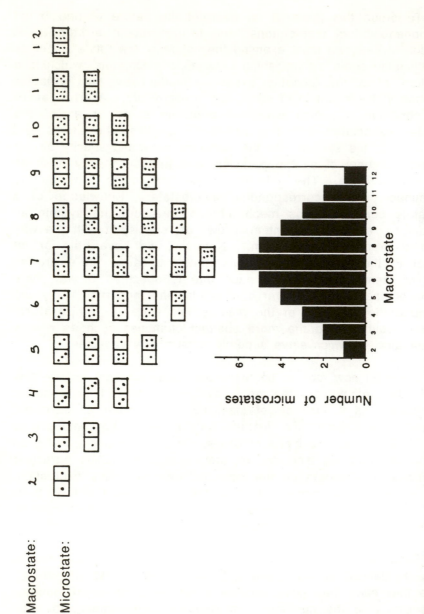

Figure 27. The combinations defining microstates of two dice, the corresponding macrostates, and the number of microstates for each.

Macrostate: 3   4   5   6   7   8   9   10   11   12   13   14   15   16   17   18

Microstate: 111 112 122 222 223 233 333 334 335 336 346 356 366 466 566 666
                 (3) (3)   (3) (3)   (3) (3) (3) (6) (6) (3) (3) (3)

                                113 123 124 134 234 235 236 246 256 266 456 556
                                (3) (6) (6) (6) (6) (6) (6) (6) (6) (3) (6) (3)

                                        114 115 125 135 136 245 156 166 446 555
                                      (3) (3) (6) (6) (6) (6) (6) (3) (3)

                                      133 116 225 145 344 255 445 455
                                      (3) (3) (3) (6) (3) (3) (3) (3)

                                          224 126 244 155 345 355
                                          (3) (6) (3) (3) (6) (3)

                                          144 226 146 444
                                          (3) (3) (6) (3)

Number of   1   3   6   10   15   21   25   27   27   25   21   15   10   6   3   1
Microstates

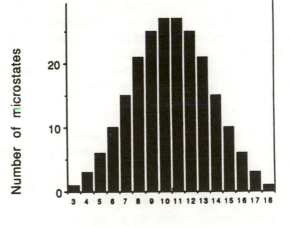

Figure 28. The microstates and macrostates for a system of three dice. The three digits correspond to the showing numbers on the dice; the numbers in parenthesis are the numbers of ways each combination can be achieved.

Thus the system of two dice has eleven macrostates with the number of microstates increasing linearly from the "lowest" state 2 to state 7 and then decreasing linearly to state 12. Suppose our system consists of three dice; how many microstates are there, how many macrostates, and how many microstates correspond to each macrostate? With three dice, there are $6 \times 6 \times 6 = 216$ possible throws and therefore 216 possible microstates. The macrostates range from 3 through 18, so there are only 16 possible macrostates. Since every microstate belongs to some macrostate, there must be several microstates corresponding to certain macrostates. Figure 28 shows the distribution.

With three dice, the distribution shows a tendency to round and develop a shape like a dome. The number of microstates increases first more rapidly than linearly, then less rapidly through the peak, and then down the other side in a symmetrical way.

As the number of dice is increased, the curve drawn through the number of microstates per macrostates becomes *increasingly peaked in the vicinity of the most probable macrostate*. For a large number of dice, the distribution of microstates among macrostates takes on the form of Figure 29, with the central peak becoming narrow *very* rapidly as the number of dice increases beyond four or five.

The narrowing of the distribution is a property of virtually all systems containing large numbers of relatively independent constituents. It is a natural result of the use of summary variables to describe the macrostate. There is a very important physical consequence of this narrowing, provided we make one assumption. *All microstates have the same probability of being observed.*

There is every reason to believe this is true, although it is still a matter of assumption. We do know that any microstate that is ever attained by almost any system will be attained again; at least microstates arbitrarily similar to the originally-selected microstate will be attained again. The demonstration of this property of microstates is called the **ergodic theorem.** Almost all systems, with the exception of some of the very simplest, are believed to conform to this theorem and are said, therefore, to be ergodic.

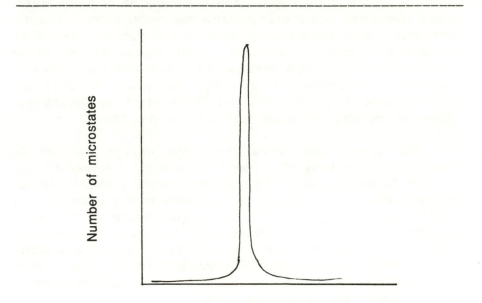

Figure 29.   Distribution of microstates among macrostates for a system of many dice.

Recognizing ergodic behavior, that systems must revisit or pass arbitrarily near every accessible microstate pulls us partly out of the dilemma of Zermelo's Paradox, but does not yet resolve the problem.   However, if we now reconcile the system's property of being ergodic with the peaked shape of the distribution of microstates, as in Figure 29, then we finally can understand how a system can only be observed in the macrostate corresponding to the equilibrium state.   Here is the argument.

For a typical system of many atomic and molecular particles, whose energy is roughly k_BT per particle at room temperature of 300°K, or 300 $\mathbb{R}$ per mole, the total number of accessible microstates is absolutely *enormous*. *And practically all of these microstates correspond to the most probable macrostate* accessible to the system or are so close to it as to be indistinguishable from it by ordinary macroscopic observations. The microstates are all so clustered around the most probable macrostate that the number of microstates *detectably* far from that macrostate is vanishingly small. There is simply almost no likelihood of our finding the system in any other macrostate than the most probable.

The system must some time, pass through each of its improbable macrostates (if it lasts long enough). However these are each extremely rare. The criterion of equal probability for all microstates says that it is just as improbable to find a system in a specific microstate that belongs to a probable macrostate as it is to find the same system in a microstate that belongs to an improbable macrostate. There are just lots and lots more microstates available for the probable macrostates. The probable average time it takes for a system to return to any microstate that it has once occupied is called the **Poincaré recurrence time**.

We can get some idea of the magnitude of Poincaré recurrence times by looking at a very simple system. Consider a deck of cards dealt out in a specific order, any order, as defining a microstate. How long should we expect to wait before a duplicate of that order is dealt again? Assume that the cards are shuffled well after every deal. There are 52 cards, so there are 52! possible microstates, all presumably equally likely. (The 52 with an exclamation is read "52-factorial", which means $52 \times 51 \times 50 \times \ldots \times 3 \times 2 \times 1$.) The number 52! is a rather large number, about 4.4×10^{66}. If we could carry out one shuffle and deal per second, it would take over 10^{59} years to accomplish 4.4×10^{66} deals. The universe itself is only about 10^{13} years old! On average, it would take about 10^{45} times the age of the universe to repeat a deal exactly.

Consequently the resolution of Zermelo's Paradox and of the

existence of equilibrium distributions for macrostates is this: any system wanders among its microstates in a way that *is* really governed by mechanics.   If that system consists of more than a few particles, almost all of its microstates correspond to its most probable macrostate, or are indistinguishably close to that most probable macrostate.   Because any even slightly complicated system is practically always in its most probable macrostate, that state is the only one we see (if the system has been left alone) and therefore we call that state *the* equilibrium state.   The distribution of velocities or quanta when the system is in its most probable state is called the equilibrium distribution.   Really, from the viewpoint of a microscopic analysis, it is the overwhelmingly large probability of this distribution that makes the corresponding macrostate the one that persists and that we observe.

This behavior gives us a deep insight into the directionality of time itself, that is, why events run only one way in time: objects break, they do not spontaneously repair themselves; gases expand, they do not spontaneously contract; we age, we do not spontaneously grow younger.   All of these are manifestations of a system evolving from one macrostate to another.

A system that changes in time moves from a macrostate of high probability to another of even higher probability, always moving into the next available macrostate having the highest probability. Even the universe itself can be thought of as evolving through a succession of distributions of increasing probability.

Problems

1. Construct the probability distribution of the macrostates of four dice and draw a graph of that distribution, in the manner of Figure 28.

2. How many ways can one get "21" with two cards in an ordinary deck? That is, how many ways can one be dealt one ace and one 10 or face card? Show that the total number of pairs of cards one can be given with an ordinary deck is $(52 \times 51)/2$, and find the probability of getting "21" with two cards. This number is low enough to make the event improbable but high enough to make the possibility tantalizing enough to be the basis of a game.

3. Construct imaginary sequences of coin-tossing results as follows:

   a) 15 sequences of 6 tosses;

   b) 15 sequences of 25 tosses.

   Then plot the frequency distribution for the number of heads among the 15 sequences, the first from zero to 6 and the second from zero to 25. The *number* of heads becomes the variable designating the macrostate and the specific sequence is the microstate.

   Now do the same again but with real sequences of coin tosses, 90 tosses in all for part a) and 375 tosses for part b), and plot the actual distributions.

   Finally, compare your results based on imaginary sequences with those from actual sequences. This problem illustrates both the tendency toward peaking around the most probable distribution as the microstate becomes more complicated (a longer sequence in this case) and also the difficulty we have in "creating" random sequences of events.

17. A MICROSCOPIC VIEW OF ENTROPY AND THE SECOND LAW OF THERMODYNAMICS

The number of microstates corresponding to a given macrostate is clearly a measure of the degree of specificity of the state, so that more ordered states (like straight flushes and 24-point bridge hands) are improbable simply because they correspond to very few microstates. We can gain considerable insight about a system if we characterize its macrostates by the number of microstates that correspond to each, but those numbers are not the most useful to express the information we want. For one thing, the numbers are too big. For another, we would like to deal with variables that change by simple arithmetic rules when the system changes state. However the number of microstates of a complex system typically changes in a complicated way when the system changes its macrostate.

For example, if the volume of a gas of n molecules triples (and the temperature remains constant), then the number of available microstates increases by a factor of 3^n, simply because each particle has three times the available places to be. This is a multiplicative relation, a factor of three for each molecule, and n of those factors. It would be convenient to have instead a quantity that characterized the change by an additive relation instead of a multiplicative one.

Suppose we have a function that behaves according to a multiplicative operation and we would rather have a second function, related to the first, that behaves according to addition. Then we know we should take the logarithm of the first function. That is, if

$$Z = xy$$

then
$$\log Z = \log x + \log y .$$

(110)

Let us call W the number of microstates corresponding to a given macrostate; there are W_1 microstates corresponding to macrostate

1. The function we want, then, is the logarithm of $\mathbb{W}$; in fact, we shall take that logarithm to the natural base e, 2.71828 and which we write as $\ln \mathbb{W}$. (When "$\ln$" is used to mean "base e", then "log" is taken to mean "base 10".) We customarily multiply this by a scale factor, which also puts dimensions on the function we are about to define. The scale factor is Boltzmann's constant k_B, per particle, or the gas constant $\mathbb{R}$, per mole. The function we now define is the **entropy**, S:

$$S = k_B \ln \mathbb{W} \quad \text{(per particle)}$$

or $\qquad\qquad S = \mathbb{R} \ln \mathbb{W} \quad \text{(per mole)}$. $\qquad\qquad$ **(111)**

Note that the dimensions of entropy and heat capacity are both heat per degree. The gas constant, the Boltzmann constant and the molar specific heat all have dimensions of heat per degree, per quantity of matter, counted either per particle or per mole and not by mass. Moreover entropy is a property that depends only on the macrostate of the system, and not on how it was prepared. We do not look to see which of its $\mathbb{W}$ microstates corresponds to the system; we do not care to know. In essence, however we prepare the system, it finds its own equilibrium state for which $S = R \ln \mathbb{W}$ is the entropy of each mole.

Now we have laid all the groundwork for a first statement of the Second Law of Thermodynamics:

The entropy S, a state function of a system, always increases in any natural process.

It is possible for the entropy of restricted subsystems to decrease, provided the *net* entropy of the subsystem and its surroundings increases. We can arrange a deck of cards, but we must do work if we want to stack our deck. Whenever a natural process seems to involve a decrease of entropy, there must be some compensating gain in the total entropy. For example, when water freezes on a cold day, its order increases, and the solid loses many of the microstates that were available to the liquid. The reason freezing of water can be spontaneous is that the entropy of the universe increases when the quanta of energy escape from the solid and go into the radiation field or into some heat sink, such as the cold surroundings.

There are many ways to state the Second Law. The others we shall consider here make no mention of the relation of entropy to microstates. It is important to realize that our first statement is founded on the hypothesis that accessible microstates of a system are all equally probable. It is also worth noting the close relation between the hypothesis of equal probability, the principle of equipartition of energy and the Second Law. When energy is equipartitioned among the degrees of freedom of a system, it is spread in just that way that corresponds to the macrostate of greatest probability *because there are more microstates corresponding to equipartition* than to any other assignment of energy in the system.

Suppose we immerse a hot object in a colder medium, and call the whole our system. As energy is carried from the hot object to the surrounding medium, the hot object cools, the medium warms and the system eventually comes to thermal equilibrium, with one common temperature. The process of conversion of energy stored as an unequal distribution of temperatures into a uniform-temperature distribution is a process of equipartition, i.e. a process that increases entropy.

It is unimaginable that the heat of the once-hot object would flow back spontaneously to it after thermal equilibrium is established. This would require that heat flow of its own accord from a cold object to a hot object. This is just as difficult to imagine as the idea that all the molecules of air in a room would concentrate spontaneously in one small part of the room. The natural, spontaneous process has heat flow from hot to cold. There is a spontaneous natural direction of evolution: even while microscopic mechanics goes on at the scale of particles, the system evolves in an unidirectional way in time. The statement that heat only flows spontaneously from a high temperature to a low temperature will become one of our ways to express the Second Law of Thermodynamics.

There is one final point to be made here, about the microscopic statement of the Second Law. This statement of the Law, in particular, opens an opportunity to find the *absolute* entropy of a single state, rather than the entropy change between states. All the other statements, referring only to macroscopic variables, are expressions about the *change* of entropy and not about its absolute amount. The Second Law alone does not tell us W ; that must come from mechanics. However it does tell us how to use the information from mechanics to deal with a many-particle system.

Problems

1. Compute the entropy for each of the macrostates of Problem 1 at the end of Chapter 16, using for Boltzmann's constant k_B 1.38×10^{-23} joule/degree K.

2. Compute the entropy for the macrostate corresponding to "21" as in Problem 2 at the end of Chapter 16. Then compute the entropy for a macrostate consisting of all pairs *except* the combinations (microstates) yielding "21". Show that the latter could be evaluated just as well by defining the macrostate as the set of all pairs, and neglecting the set corresponding to "21".

18. THE THERMAL DEFINITION OF ENTROPY AND MACROSCOPIC STATEMENT OF THE SECOND LAW OF THERMODYNAMICS

Returning to the macroscopic viewpoint, we present an apparently altogether different definition of entropy and restate the Second Law, using only the macroscopic concepts of heat, work and temperature. As with energy, defining entropy in terms of purely macroscopic variables leaves the quantity specified only to within an additive constant. That is, we can define only *changes* in entropy, not absolute entropy itself, so long as we restrict ourselves to defining it and stating the Second Law solely in terms of macroscopic state variables. Let us start by considering two states of a system, A and B, *having the same temperature*. Then we define the change in entropy ΔS as

$$\Delta S = S_B - S_A = Q_{reversible}/T \; . \qquad (112)$$

The entropy change is defined as the heat absorbed in a *reversible* process connecting states B and A, per degree of temperature. Entropy changes are not defined as the actual heat absorbed, per degree, but as the limiting value of heat absorbed per degree in the process in which the system comes essentially to equilibrium at every tiny step along the way.

The entropy so defined has the same dimensions as the heat capacity. Here, in contrast to the definition $S = k_B \ln \mathbb{W}$ of Chapter 17, the dimensions are inherent in the definition. The dimensions imposed by the previous definition seem arbitrarily fixed by the choice of the coefficient, either k_B or R. Entropy, according to our new definition, represents the energy absorbed as *heat* – – energy distributed among the degrees of freedom, per degree of temperature – – which is, in effect, the measure of the average energy in each degree of freedom. The heat $Q_{reversible}$ (or Q_{rev}, for convenience) must be the *maximum* Q absorbed, because it occurs in a process in which *all* the degrees of freedom come to equilibrium. Therefore, when heat is being absorbed, these degrees of freedom absorb as much as they need to bring the temperature in every degree of freedom to T. The distribution of energy in every degree of freedom is a Boltzmann distribution corresponding to the same temperature T. If the process is irreversible, then not all the degrees of freedom have time to respond, so they cannot absorb all the heat that they would take in if the process were arbitrarily slow.

One example in which a system cannot respond reversibly to an input of energy is diving badly into a swimming pool and landing in a belly flop. A more thermodynamic example in which some irreversibility is clearly related to the limited capability of gas molecules to absorb energy rapidly is in the response of a molecular gas to a shock wave, such as a sonic boom. The translational and rotational degrees of freedom share energy in virtually every collision, but vibrations are not so easily convinced to take on energy. Between 10 and 10^6 collisions are required for vibrations to come to equilibrium with translations and rotations; the number depends on the temperature and the particular gas.

If state A and state B have different temperatures, then there are two common ways to deal quantitatively with the transformation of the system from one to the other and, in particular, to evaluate the change of entropy. One way is to choose any convenient reversible path, and break it into steps in which the temperature is almost constant, and then add up the contributions from these steps. We number the steps 1, 2, . . . n, and suppose $Q_{rev}^{(1)}$ is absorbed in the first step at T_1, $Q_{rev}^{(2)}$ is absorbed at T_2 in the second step and so on, to obtain

$$\Delta S = S_B - S_A = Q_{rev}^{(1)}/T_1 + Q_{rev}^{(2)}/T_2 + \ldots + Q_{rev}^{(n)}/T_n \quad . \textbf{(113)}$$

The other way is to recognize that two states can always be connected by one isothermal reversible step, with T constant, and one adiabatic reversible step, in which $Q = Q_{rev} = 0$, so

$$\Delta S = Q_{rev,isothermal}/T_{isothermal} + 0 \quad . \qquad \textbf{(114)}$$

Either definition is suitable and allows us to determine entropy changes between any two states.

This statement says that a system can only make perfect use of all its heat capacity by going through the ideal limiting process we call a reversible process. It is a statement about the effect of the process on the system itself and how it can respond. An alternative and equivalent statement can be made in terms of the effect of the system on its surroundings, via the work done by the system. Recall that ΔE is a function only of state and $\Delta E = Q + W$, with Q the heat absorbed by the system and W the work done *on* the system, so $(-W)$ is the work done *by* the system if it is considered as an engine. Then if Q is less than Q_{rev}, the real work $(-W)$ done by the system on its surroundings must be less than the ideal limiting value of the work $(-W_{rev})$, which would be the *maximum* work for a transformation between states A and B. Thus, we can say that $(-W) < (-W_{rev})$. However one can make a far stronger, more precise, statement by examining the *efficiency* of a process, or of any machine that converts heat into work.

Entropy is a function of state, as its microscopic definition, $k_B \ln \mathcal{W}$, suggests. This is true despite the fact that the macroscopic definition requires that we use a process (of the special reversible type) to evaluate it. It means that the entropy change of a system undergoing a closed-cycle process must be *zero*, per cycle. We shall use this property shortly, in dealing with the concept of efficiency.

The Second Law has at least two other alternative statements, from a macroscopic standpoint. One is the statement that we saw in passing, at the end of Chapter 17:

Heat cannot be made to go from a low temperature to a high temperature without work being done.

The other is the statement:

A heat engine operating between two temperatures T_H and T_L ($T_H > T_L$) cannot convert all the heat absorbed at T_H into work; some must be exhausted as heat at T_L.

An example will illustrate our macroscopic definition of entropy and connect it to the microscopic picture. In a container with volume V_A, there are n moles of gas at temperature T_A. The box is connected to another box, evacuated, with the same volume and with walls at the same temperature. We allow the n moles of gas to expand to fill the entire volume $2V_A$. This can be done in many ways; two extremes are shown in Figure 30.

(a)

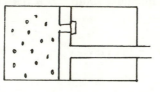 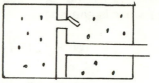

Initial state (i)   Final state (ii)

(b)

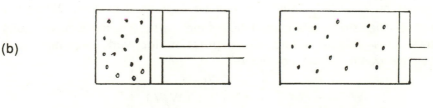

Initial state (iii),   Final state (iv),
identical to   (i)   identical to (ii)

Figure 30. Expansion of an ideal gas from V_A to $2V_A$.
    (a)   Free expansion initially as (i) and finally as (ii); $\Delta E = 0$ and $Q = W = 0$.
    (b)   Reversible expansion, $\Delta E = 0$ so $(-W) = Q > 0$ and $Q + W = 0$; that is, work is done and so heat must be absorbed.

The process shown in (a), the free expansion, is one we have already seen. Because the process is isothermal and the gas is ideal, its energy E depends only on T. We immediately infer that $\Delta E = 0$, so the heat absorbed is equal to the work done, i.e. $Q = (-W)$, from the First Law. But no work is done in a free expansion because the external pressure is zero. Hence $Q = (-W) = 0$. Yet some property of the gas much have changed, besides the volume. What is it?

To answer that question, we consider the reversible isothermal expansion from (iii) to (iv): $\Delta E = 0$, as in the free expansion, and $Q = (-W)$ but neither Q nor W is zero now. In fact the work done on the system is expressible as the sum of the increments $-P_{external}\Delta V = -p\Delta V$ (because of reversibility). But the pressure p is governed by the ideal gas law in this system: $p = n\mathbb{R}T/V$. Therefore the work is the sum

$$W = [-n\mathbb{R}T\Delta V_1/V_1 - n\mathbb{R}T\Delta V_2/V_2 - ... - n\mathbb{R}T\Delta V_{last}/V_{final}]$$

$$= -n\mathbb{R}T\Sigma(\Delta V_j/V_j) , \qquad (115)$$

where j is just an index number, for the general j^{th} step. The quantity is a constant, $n\mathbb{R}T$, times the sum of all the fractional changes of V, $\Delta V_j/V_j$. This sum can be expressed exactly. When the sizes of the steps becomes infinitesimal, and the number of steps grows large without bound, the expression for work becomes (in a result we quote but do not derive)

$$W = - \lim_{\text{number of steps}\to\infty} \left\{ n\mathbb{R}T \sum_{V_i}^{V_f} (\Delta V_j/V_j) \right\}$$

$$= -n\mathbb{R}T \, \ell n(V_f/V_i) . \qquad (116)$$

(We let the indices i and f, below and above the summation sign Σ or sigma, indicate "initial" and "final" meaning that all the steps should be included in the sum, as in Eq (115).) The first line uses a conventional notation to indicate that we get the result of interest by taking the limit indicated under the "lim" expression. The heat taken in must be equal to the work done by the system so Q also satisfies this condition:

$$Q = nRT\ln(V_f/V_i) .$$

The entropy change ΔS for the reversible, isothermal process is therefore

$$Q/T = Q_{rev}/T = nR\ln(V_f/V_i) , \qquad (117)$$

or $Nk\ln(V_f/V_i)$ (where N, as before is the number of particles), or

$$k\ln(V_f/V_i)^N .$$

When $V_f > V_i$, we know from everyday experience that the process is spontaneous. But this is exactly the case that makes $\ln(V_f/V_i)$ positive. Hence we conclude that the commonplace observation that isothermal expansions are spontaneous implies that they are associated with positive entropy changes. When $V_f < V_i$, the process will *not* occur unless there is some outside work done, and ΔS of the system is necessarily negative.

For the isothermal free expansion, we said Q = 0, but for the reversible expansion, $Q = nRT\ln(V_f/V_i)$. We used the latter, not the former, to evaluate ΔS. Why? The former is *not* a reversible process, so $W > W_{max}$, $Q < Q_{max}$ and Q/T is *less* than that for the reversible process on which the definition is based.

Problems

1. Evaluate the entropy change due solely to the isothermal expansion of one mole of a fluid, from an initial density of about that of liquid water, 1 mole per 18 cm^3, to a final density of 1 mole per 22.4 liters, at 1 atm pressure and temperature of 273°K. This quantity is about the same for many, many materials <u>because</u> atoms and small molecules are all about the same size and all gases at 1 atm pressure and 273°K are relatively ideal. Explain the underlined "because". (The fact that this entropy is roughly the same for many substances is exemplified in **Trouton's Rule**, which states that the reversible heat per degree, per mole, to vaporize a liquid at its boiling point is approximately 21 calories/mole degree.)

2a. What is the entropy change when one mole of an ideal, monatomic gas is heated reversibly at constant volume from temperature T_1 to temperature T_2? (Hint: how is the heat absorbed related to the temperature change of a gas? Examine the incremental changes in a manner analogous to the volume changes discussed in the text.)

 b. If the heating is done irreversibly how does the entropy change differ from that of part (a)?

 c. Suppose the gas is heated irreversibly so that it absorbs as much heat as the gas does in process (a) above. How do the two final states differ?

3. Any two states of a system of one substance can be connected by two reversible steps, one at constant temperature and the other, adiabatic.

 a. Sketch an indicator diagram to show how this is possible for the two examples given below.

b. Show that the entropy change between two states of an ideal gas is always of the form

$$\Delta S = nR \ln V_f/V_i + nC_V \ln T_f/T_i \ .$$

(You may use the results of Problem 2.)

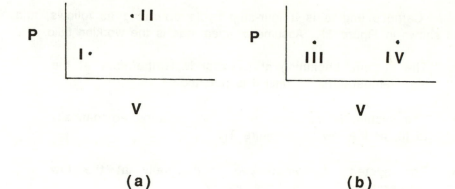

(a)   (b)

19. THE CARNOT ENGINE AND EFFICIENCY

A remarkable thing about the conversion of energy from heat, its most equipartitioned form, into *work*, is the requirement that heat must be transferred from a hot reservoir to a cold reservoir. We cannot run a heat engine through a closed cycle at a single temperature. Sadi Carnot (1796–1832) realized this when, in 1824, he proposed his formulation of what we now call the Second Law, and his famous engine.

Carnot's engine is a four-step cycle operating as follows, and as shown in Figure 31. Assume an ideal gas is the working fluid.

1. The system, beginning at I, expands isothermally at the high temperature T_H until II is reached.

2. The system is insulated at II, and expands adiabatically to III, at the low temperature T_L.

3. The system is compressed isothermally at the low temperature T_L, until it reaches IV.

4. The system is compressed adiabatically from IV back to I.

This is not an engine anyone wants to build: it would require very large changes of volume and would be terribly unwieldy compared with engines we actually use. However, it is *the* pristine model for inferring the best performance any real engine can approach.

We need not worry here about such questions as how IV is selected to assure the return to I. Let us concentrate on finding how this system operates. We shall want to find the work per cycle, the heat taken in, the energy change per cycle and the entropy change per cycle. We shall need these because we want to examine the efficiency η of the system:

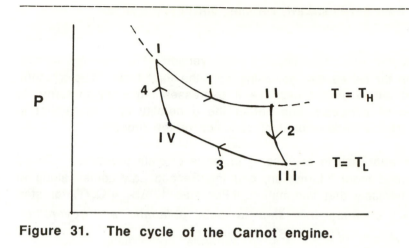

Figure 31. The cycle of the Carnot engine.

$$\eta = \text{net work out, per cycle/heat in, per cycle} \qquad (118)$$

The net work in a cycle is $W_1 + W_2 + W_3 + W_4$. The *total* heat is the same as the total work, since $\Delta E = 0$. Moreover the work done *by* the system in step 2 exactly cancels the work done *on* the system in step 4:

$$W_2 = \Delta E_2 \text{ because } Q_2 = 0$$

and similarly for W_4. But recall that the only energy an ideal gas has is the kinetic energy of its non-interacting particles, so that E depends only on T for an ideal gas. Hence $(-W_{net})$, the work done *by* the system in step 2 is equal to the work done *on* the system in step 4 because both steps operate between the same temperatures with $Q = 0$. Thus the net work done per cycle is

$$-W_{net} = -W_1 - W_3 \; .$$

The heat *absorbed* is, of course, Q_1. Hence the efficiency is $-(W_1 + W_3)/Q_1$. However $(-W_1) = Q_1$, and $(-W_3) = Q_3$ because these steps are isothermal, so $-(W_1 + W_3) = Q_1 + Q_3$. Thus from the identity of the gas and from the First Law, we infer that

$$\eta = (Q_1 + Q_3)/Q_1 \text{ , with Q always Q}_{rev} . \qquad \textbf{(119)}$$

This is as far as we can go, using only the First Law. It is a useful statement but is limited because it expresses efficiency in terms of properties of *process*, and not in the generality of a statement in terms of *states*, independent of specifics of the process.

We want now to express η in terms of state properties. To do this we need more information, and the Second Law comes along to provide precisely that information. For step 1, $\Delta S_1 = Q_1/T$; for step 2, $\Delta S_2 = 0$ *because this step is both adiabatic and reversible.* Similarly, $\Delta S_3 = Q_3/T$ and $\Delta S_4 = 0$.

The Second Law assures that

$$\Delta S_1 + \Delta S_2 + \Delta S_3 + \Delta S_4 = 0 . \qquad \textbf{(120)}$$

Therefore $\qquad \Delta S_1 + \Delta S_3 = Q_1/T_1 + Q_3/T_3 = 0 , \qquad \textbf{(121)}$

and we can say now that $Q_3 = -Q_1(T_3/T_1)$. The heat *deposited* at T_3 is opposite in sign to Q_1, and is directly proportional to the ratio of low-to-high temperatures. The maximum fixed by the Second Law thus relates Q_1, the amount of heat absorbed, to Q_3, the amount of heat released at the temperature of the cooler reservoir.

Now we substitute for Q_3 into the expression **(118)** for efficiency and we have the form we want:

$$\eta = (Q_1 + Q_3)/Q_1 = [Q_1 + (-Q_1 T_3/T_1)]/Q_1 ;$$

cancel Q_1's so $\qquad \eta = 1 - (T_3/T_1) \text{ or } (T_1 - T_3)/T_3 . \qquad \textbf{(122)}$

This expression is the one we sought. It tells us precisely how effectively Carnot's ideal engine can convert heat to work.

This derivation leads us to more information. We can infer that no engine operating between T_H and T_L can be more efficient than the Carnot engine, by calling once more upon the Second Law. Let us suppose a *reversible* engine X, call it the Tonrac engine, is more efficient than the Carnot engine. Then X would use only an amount of heat $Q'_1 < Q_1$, the Carnot heat input, to produce the same work (–W) that the Carnot engine produces. If this is the case, we can use the Tonrac engine to run the Carnot engine *backward*, opposite the direction of the arrows in Figure 31. This process takes heat Q_3 out of a cold reservoir at T_L and dumps heat Q_1 at the high temperature T_H. The net work done by the two engines, the Tonrac and the Carnot, is zero, because they are equal and opposite. The energy change after one cycle is zero. But we have taken Q'_1 out of the hot reservoir at T_H and have put Q_1 back into the same reservoir! We know $Q_1 - Q'_1$ is positive, so we must have taken heat $(Q_1 - Q'_1)$ from a low-temperature reservoir and converted it to heat at a higher temperature with no net expenditure of energy. This is a clear violation of the Second Law. The only flaw in the logic is the assumption that the Tonrac engine X exists. Since its existence is the only postulate or logical step we can question, we must discard that supposition. We are left with the statement that no engine operating between T_H and T_L can be more efficient than the Carnot engine. Carnot's device is an ideal limit, a boundary on how well heat can be converted into work.

The argument we just used can be turned around to show that no *ideal reversible* engine cannot operate below the Carnot efficiency either. Suppose we have an ideal engine Y that is less efficient than the Carnot engine. Again, we select a version of Y that produces the same work per cycle as the Carnot engine. This time we drive the Y engine with the Carnot engine The net work is zero. The Carnot engine draws heat Q_1 from the hotter reservoir, and Y deposits heat Q''_1 into the same reservoir, with each cycle. The deposited heat Q''_1 must be larger than Q_1 because Y is less efficient than the Carnot engine but, run forward, produces the same work.

Hence we could violate the Second Law if there were an ideal engine Y that was less efficient than the Carnot engine. The conclusion is that all ideal engines operating between temperatures T_H and T_L have the same efficiency η, equal to $(T_H - T_L)/T_H$.

The nonideal engine is not as efficient as $(T_H - T_L)/T_H$. How can this be? How can any engine fail to comply with the arguments we just gave to show the universality of the Carnot efficiency? The answer is again in the Second Law. If the process is irreversible, we can only require that ΔS be greater than the sum of $\Delta Q/T$. But ΔS is definitely zero for the system going around the cycle, so heat must be lost from the system in an irreversible process. Suppose, for example, that the adiabatic steps, for which $Q = 0$, are also **isentropic**, i.e. $\Delta S = 0$ along these steps. As with ideal engines,

$$\Delta S_{cycle} = \Delta S_1 + \Delta S_2 + \Delta S_3 + \Delta S_4 \qquad (123)$$

$$= \Delta S_1 + \Delta S_3 = 0 \ . \qquad (124)$$

This immediately shows that heat is lost in an irreversible cycle. Why?

For an irreversible process, e.g. step 1,

$$\Delta S_1 = Q_{1,rev}/T_H > Q_1/T_H \qquad (125)$$

and
$$\Delta S_3 = Q_{3,rev}/T_L > Q_3/T_L \ . \qquad (126)$$

Therefore the condition that entropy is greater than the sum of heat increments, per degree, tells us that

$$0 = \Delta S_1 + \Delta S_3 = Q_{1,rev}/T_H + Q_{3,rev}/T_L > Q_1/T_H + Q_3/T_L \ . (127)$$

We now relate Q_1 and Q_3 as before, except that they are connected by an inequality:

$$Q_3 < -(T_L/T_H)Q_1 \qquad . \qquad\qquad (128)$$

The efficiency remains

$$\eta = (Q_1 + Q_3)/Q_1$$

because this form depends only on its definition and the First Law. But now we must substitute from an *inequality*, to obtain the efficiency in terms of temperatures:

$$\eta < [Q_1 - Q_1(T_L/T_H)]/Q_1 \ ,$$

or

$$\eta < 1 - T_L/T_H$$

or

$$\eta < (T_H - T_L)/T_H \ .$$

Thus the *nonideal* engine has an efficiency lower than that of the Carnot engine.

Problems

1. A real engine operates between 400°K and 1000°K, at an efficiency of 40%. How well does this engine perform, relative to the ideal limit?

2. Explain, in terms of efficiency, the reason why coolers are often installed on the exhausts of large engines. What happens to efficiency as the temperature T_L goes toward absolute zero, i.e. 0°K?

3. The reversible engine Z follows the path shown inside the Carnot path:

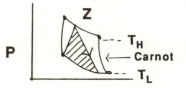

    The shaded area is the work per cycle done by Z. Both Z and the Carnot engine operate between T_H and T_L. Clearly $W_{net}(Z)$ is less than W_{net}(Carnot) so it appears that η_Z is less than η_{Carnot}. Resolve this apparent paradox.

4. Explain how the efficiency of a real Carnot-like engine that has reversible isothermal steps but irreversible adiabatic steps works out to be less than that of the Carnot engine.

20. FREE ENERGIES AND CRITERIA OF MERIT

With the development of the Second Law, thermodynamics confronted the general question of identifying the natural direction of change for *all* systems and processes. The entropy of the system and its surroundings becomes the key to answering this question. Much of the effort in late 19th and 20th century thermodynamics was concerned with rephrasing the criterion of increasing entropy of the system and its surroundings in terms of new properties of the system alone. The ideas evolved into the invention of **thermo-dynamic potentials**, quantities analogous to potentials in mechanics and electricity, whose changes are equal to the best possible performance of a system or process, either the work the system can do or the heat it can transport. The new properties are the **Helmholtz** and **Gibbs free energies**, $\mathcal{F}$ and $\mathcal{G}$, which are the first subjects to be treated in this Chapter. These are the most generally used of the thermodynamic potentials. They are functions of state that put bounds on a process variable, the amount of work that a process can do when it operates under particular constraints. The choice of thermodynamic potential depends on the constraints. Availability, which we shall examine in the next Chapter, is a closely related function but it is not a property of the system alone; it refers to external conditions that the system can approach.

Efficiency was described in Chapter 19 as a measure of performance of a system, the fraction of the energy input that is converted into useful work. Efficiency is the first of the **criteria of merit** that we can derive from thermodynamics for the purpose of comparing a real system with its ideal limit. Among these criteria of merit, this chapter will examine the **effectiveness**, the coefficient of refrigeration, and the ratio of actual to ideal power.

FREE ENERGIES

The Helmholtz free energy $\mathcal{F}$ is defined in terms of the state functions of internal energy E, temperature T and entropy S, and is therefore a state function itself:

$$\mathscr{F} = E - TS . \qquad (129)$$

The Helmholtz free energy and Gibbs free energy as well are similar to the enthalpy in being defined in absolute terms, not in differential terms like the statement of the First Law. This means that changes in free energy are simply related to *process* variables of heat and work only under those special circumstances when process variables are equal to changes of state variables. One condition required for both the Helmholtz and Gibbs free energies to be work potentials is constancy of temperature so that

$$dQ_{rev} = TdS . \qquad (130)$$

This way, a small change in $\mathscr{F}$ takes the form

$$d\mathscr{F} = dE - d(TS)$$

$$= dE - TdS - SdT$$

$$= dE - TdS \qquad \text{when T is constant}$$

$$= dE - dQ_{rev} . \qquad (131)$$

Now we can use the First Law to connect a change dE in the state variable E with the changes in process variables dQ and dW, so

$$d\mathscr{F} = dQ + dW - dQ_{rev} . \qquad (132)$$

If the process is reversible, $dQ = dQ_{rev}$ and $dW = dW_{rev}$ so

$$d\mathscr{F} = dW_{rev} . \qquad (133)$$

Therefore $\qquad \Delta\mathscr{F} = \int dW_{rev} = W_{rev} \qquad (134)$

for a *reversible*, *isothermal* process. In other words, the change in the Helmholtz free energy, $\Delta\mathscr{F}$, is equal to the minimum work required to make the system operate, or to the negative of the maximum reversible work that a system can perform when it undergoes an isothermal process. For this reason, $\mathcal{F}$ is sometimes called the work function (in German "Arbeitsfunction", and therefore sometimes represented as $\mathcal{A}$, but we reserve $\mathcal{A}$ for availability, introduced in the next Chapter).

The Gibbs free energy, $\mathcal{G}$, bears the same relation to the Helmholtz free energy as enthalpy does to energy. The Gibbs free energy is

$$\mathcal{G} = H - TS$$

$$= E + pV - TS . \tag{135}$$

Hence for a constant-temperature process,

$$d\mathcal{G} = dE + d(pV) - dQ_{rev} . \tag{136}$$

Now restrict the process to constant pressure:

$$d\mathcal{G} = dE + pdV - dQ_{rev}$$

$$= (dQ - dQ_{rev}) + (dW + pdV) . \tag{137}$$

If the process is reversible, the terms in the first parentheses cancel. The second parenthesis gives the reversible nonmechanical work $dW_{rev} + pdV = -p_{ext}dV + dW_{other} + pdV$, which includes dW_{other}, whatever electrical, magnetic and any other work against forces different from that of the surrounding atmosphere. Hence the Gibbs free energy is useful to describe the electrical work done by batteries operating at constant temperature, for example:

$$dW_{rev} = -pdV + dW_{nonmech, \, rev} \tag{138}$$

so

$$(d\mathcal{G})_{T \, const, \, revers \, proc} = dW_{nonmech, \, rev} \tag{139}$$

or

$$\Delta\mathcal{G} = W_{nonmech, \, rev} \tag{140}$$

for a reversible, isothermal process. Suppose this isothermal, nonmechanical work is the reversible work of moving a total charge of n Faradays (n moles of 1-electron charges) through a change in potential (voltage) $\mathcal{E}$ (a mole of charge is symbolized by $\mathbb{F}$ and is expressed in units of charge such as the Coulomb or the electronic charge.) Therefore the work is $n\mathbb{F}\mathcal{E}$ so

$$\Delta\mathcal{G}_{elec} = -n\mathbb{F}\mathcal{E} \ . \tag{141}$$

This is the negative of the maximum isothermal work that can be done by a battery whose chemical reaction involves the change $\Delta\mathcal{G}_{elec}$ of Gibbs free energy.

How are the changes in free energies related to the entropy changes of the system and its surroundings? The Helmholtz free energy has finite changes of

$$\Delta\mathcal{F} = \Delta E - \Delta(TS) \tag{142}$$

and for an isothermal process,

$$\Delta\mathcal{F} = \int dQ + \int d W - T\Delta S \ . \tag{143}$$

If the system does no nonmechanical work and *remains at constant volume*, then $\int dW = 0$, and

$$\Delta\mathcal{F} \text{ becomes } \int dQ - T\Delta S \ .$$

Now we examine the meaning of dQ from the viewpoint of the surroundings with which this heat is being exchanged. The surroundings consist only of a heat reservoir, and therefore can change entropy *only* through exchanges of heat. This implies that

$$dQ = -TdS_{surroundings} \tag{144}$$

with a negative sign since a flow of heat out of the system is a flow of heat into the surroundings. This means that for an isothermal process in which the volume of the system is constant and no other work is done,

$$\Delta \mathcal{F} = - T\Delta S_{surroundings} - T\Delta S$$

$$= - T(\Delta S_{surroundings} + \Delta S_{system}) \qquad (145)$$

(we have subscripted ΔS for emphasis) or

$$\Delta \mathcal{F} = - T\Delta S_{"universe"} \cdot \qquad (146)$$

Thus $-\Delta \mathcal{F}/T$ is equal to the change of entropy of the system plus surroundings, or the entropy change of the "universe" under the restricted conditions just described. Note that this does *not* require reversibility. If the process is reversible, then $dQ = T\Delta S$ and the special condition applies that

$$\Delta \mathcal{F} = 0 , \qquad (147)$$

for a *reversible, isothermal* process at constant volume in which no work is done.

The analogous line of reasoning for $\Delta \mathcal{G}$ holds but for processes at constant pressure. The condition that no nonmechanical work be done means that

$$dW = - PdV \qquad (148)$$

and if the process is mechanically reversible, so $p = P$, $dW = -pdV$

so
$$d\mathcal{G} = dE + d(pV) - d(TS) \qquad (149)$$

for a mechanically reversible process becomes
$$d\mathcal{G} = dQ - pdV + pdV + Vdp - TdS - SdT$$

$$= dQ + Vdp - TdS - SdT . \qquad (150)$$

When p and T are constant,

$$d\mathcal{G} = dQ - TdS$$

$$= -T(dS_{surroundings} + dS_{system})$$

$$= -TdS_{universe} \cdot \qquad\qquad (151)$$

Thus $\qquad\qquad \Delta\mathcal{G} = -T\Delta S_{universe} \qquad\qquad (152)$

for an isothermal, isobaric process in which no other work is done. For a reversible process satisfying these conditions,

$$\Delta\mathcal{G} = 0 \cdot \qquad\qquad (153)$$

The entropy change $\Delta S_{universe}$ is always positive in any naturally-occurring process. Thus Eqs. **(146)** and **(152)** tell us that the Helmholtz and Gibbs free energies must always *decrease* – – under the conditions in which **(140)** and **(146)** are valid – – in any spontaneous change. The conditions of **(147)** and **(153)** are the conditions for neither a spontaneous change nor an impossible process, i.e. the conditions for equilibrium.

This gives us an opportunity to glimpse another powerful aspect of the generality of thermodynamics. In particular, thermo-dynamics tells us about the equilibrium between a liquid and a solid, a liquid and a vapor, or any two *phases* of a material that may coexist. When a pure liquid stands in equilibrium with its vapor at a fixed temperature, the pressure of that vapor is constant and the transfer of material from liquid to vapor or from vapor to liquid must conform to **(153)**. This means that if material is transformed from one phase to another, the Gibbs free energy does not change in the process, or, in other words, the *Gibbs free energy* of any unit of material is the same in *all the phases in equilibrium*. In terms of an equation,

$$\mathcal{G}_{liquid,\ per\ unit\ material} = \mathcal{G}_{gas,\ per\ unit\ material} \qquad (154)$$

for a liquid and a gas in equilibrium.

Recall that the condition for one pure phase to be in equilibrium is that its thermodynamic variables satisfy its equation of state, meaning that of p, V and T, only two are independent. Thus every equilibrium state of a pure phase corresponds to a point in a pressure-volume space. The conditions that two pure phases be in equilibrium with each other are that the material satisfy two equations of state simultaneously, so that *two* of the three variables, p, V and T are determined when one other is specified. If we fix T, then p and V are determined; if we fix V, then p and T are determined. In a plot of pressure versus volume, the states corresponding to two phases in equilibrium lie on – – or form – – a curve, not an area. We say the two phases coexist on that curve. Eq. **(153)** is another statement of the physical relationship that must hold along the coexistence curve for any two pure phases. The relation between the compatibility of two equations of state and Eq. **(153)** is contained in the conditions that a) two subsystems (e.g. two phases) in equilibrium must be at a common temperature and b) two systems in equilibrium with each other must exhibit no difference in the pressures they exert on one another. The two equations of state must give common temperatures and pressures. Finally c) equilibrium of two open systems implies that a transfer of material can be made *reversibly* from one phase to the other. Hence the conditions for equilibrium of two phases are the same as the conditions that there is no change in $\mathcal{G}$ if we transfer material between the phases. If the transfer of material is spontaneous, then it must be ordinarily be associated with a decrease in $\mathcal{F}$ or $\mathcal{G}$, depending on whether volume or pressure is constant. (One can construct conditions with other variables constant under which less common potentials must be used.)

We have introduced $\mathcal{F}$ and $\mathcal{G}$ because they represent maxima in the work a system can do under special but frequently-met circumstances. We have examined the relationship of $\Delta\mathcal{F}$ and $\Delta\mathcal{G}$ to $\Delta S_{universe}$ to show how the free energies, being equal to the total entropy change, have natural directions of change, under particular circumstances. And, we have illustrated the physical meaning of $\Delta\mathcal{G}$ for two cases, electrical work and phase equilibrium.

The Helmholtz and Gibbs free energies are by no means the only state functions that serve as limits to the work that can be done by a system. They are however by far the most important because they limit the work that can be done under the most commonly-met constraints. It is possible to construct other potentials that limit the work a system can do under any sets of constraints that permit the limiting process to be reversible. It is even possible to construct potentials whose changes are equal to the maximum work a system can do in a *fixed finite time*, for any set of constraints that allow temperature, pressure and concentration to be well-defined in every small but macroscopic volume in the system. The generalized potentials for reversible processes and the potentials for processes at finite rates lie beyond the scope of this text, so we shall not pursue them except to indicate in the next chapter how they can be used to help define criteria of merit.

ENERGY AND FREE ENERGY CHANGES IN CHEMICAL PROCESSES

Thus far, our analysis has dealt only with changes of pure substances, so that the energy, enthalpy and free energies have depended on pressure and temperature (or volume) but since the composition has been constant in all those examples, it does not appear in any of the expressions. Now suppose the system contains more than one phase or more than one substance, or both. The internal energy and all the other state functions depend on how much there is of each form of matter. We express this by introducing the chemical potentials, conventionally designated for the ith species or phase by μ_i. **Chemical potentials** are nothing more than the rate at which the internal energy changes with an increase in the amount of the corresponding species. If n_i is the number of moles of species i, and dn_i is the change in that amount, whether added from the outside or generated by a chemical reaction, the energy E is changed by the small amount dE as a result of adding dn_i. The chemical potential μ_i is defined to be

$$\mu_i = dE/dn_i \, , \tag{155}$$

with all the other n_j's, as well as pressure and temperature, held fixed.

For a system with s different chemical components that can undergo chemical and physical changes, any small reversible change ΔE in the total energy (with only mechanical work) can be written in terms of the changes ΔS and ΔV in entropy and volume, and Δn_1, Δn_2, ... Δn_s in amounts of each of the s species:

$$\Delta E = T\Delta S - p\Delta V + \mu_1 \Delta n_1 + \ldots + \mu_s \Delta n_s . \qquad (156)$$

One example we already discussed is the system of a liquid and its vapor. Here, μ_1 and μ_2 represent the chemical potentials μ_{liq} and μ_{vap} of the liquid and vapor phases. Suppose the system is closed so that the total mass of liquid plus vapor is fixed; then if the amount of liquid increases, the amount of vapor must decrease exactly the same amount. In terms of an equation, $\Delta n_{liq} = - \Delta n_{vap}$. A change in E for a transformation of Δn_{liq} moles from vapor to liquid is therefore

$$\Delta E = T\Delta S - p\Delta V + (\mu_{liq} - \mu_{vap})\Delta n_{liq}$$

$$= T\Delta S - p(V_{liq} - V_{vap}) + (\mu_{liq} - \mu_{vap})\Delta n_{liq} \qquad (157)$$

Since the volume of liquid is ordinarily very much smaller than the volume of the same material in the vapor, ΔE can often be approximated closely by replacing $p(V_{liq} - V_{vap})$ with $-pV_{vap}$, neglecting the term pV_{liq}. Moreover ΔS, the entropy change, is usually approximately equal to $\mathbb{R} \ln(V_{liq}/V_{vap})$ (recall "$\ln$" is the logarithm to the natural base e), when the final state is liquid and the initial state is vapor. The lost volume of vapor can be estimated from the ideal gas law as $V_{vap} = \Delta n_{liq}RT/p$ and the newly added volume of liquid as $\Delta n_{liq}\mathbb{V}$ where we let $\mathbb{V}$ be the volume of one mole of the material in any condensed phase. The result is that

$$\ell n(V_{liq}/V_{vap}) = \ell n[\Delta n_{liq}\mathbb{V}/(\Delta n_{liq}RT/p)] = \ell n[p\mathbb{V}/RT] ,$$

so
$$\Delta E \approx RT\ell n(p\mathbb{V}/RT) + pV_{vap} + (\mu_{liq} - \mu_{vap})\Delta n_{liq} . \qquad (158)$$

The changes of Helmholtz and Gibbs free energies for chemical changes are easy to obtain for reversible processes, for which $dQ = TdS$ and $dW = -pdV + dW_{nonmech}$. Since

$$\mathcal{F} = E - TS \qquad (129)$$

and
$$\mathcal{G} = E + pV - TS , \qquad (135)$$

their infinitesimal changes in a reversible process are

$$d\mathcal{F} = TdS - pdV + dW_{nonmech} - TdS - SdT + \sum_{i=1}^{s} \mu_i dn_i$$

$$= -pdV + dW_{nonmech} - SdT + \sum_{i=1}^{s} \mu_i dn_i . \qquad (159)$$

and
$$d\mathcal{G} = Vdp + dW_{nonmech} - SdT + \sum_{i=1}^{s} \mu_i dn_i . \qquad (160)$$

The vapor-to-liquid example now becomes especially cogent because, if the liquid and vapor are in equilibrium, the pressures exerted by the two phases are equal so $dp = 0$, the temperature is constant so $dT = 0$ and no work other than pressure-volume work is done, so in this case

$$d\mathcal{G} = (\mu_{liq} - \mu_{vap})dn_{liq} . \qquad (161)$$

But the previous section showed that the Gibbs free energy is the same for one mole in the two phases. Thus $d\mathcal{G} = 0$ and

$$\mu_{liq} = \mu_{vap} \equiv \mu . \qquad (162)$$

This is one example that illustrates the general rule, which we shall state but not derive here:

> When two or more chemically or physically inter-
> convertible states of matter are in contact and in
> equilibrium, the chemical potentials of all the
> components are equal.

This rule is an extension to chemical changes of the conditions that the temperatures and pressures of subsystems in equilibrium with each other must be the same. The chemical potentials μ_i are intensive variables analogous to temperature and pressure: the amounts n_i of each species play the roles analogous to the extensive variables of entropy and volume.

Problems

1. Consider the chemical equilibrium between two chemical isomers, two chemical species with the same atomic composition but with the atoms arranged differently. For an example think of hydrogen cyanide, HCN, and isocyanic acid, HNC, both of which are gases. Show that in chemical equilibrium between HCN and HNC,

    $$HCN \leftrightarrow HNC \quad ,$$

    the chemical potentials of the two species are the same. Construct your argument parallel to that used to demonstrate the equality of chemical potentials of a liquid and a vapor of the same material in equilibrium.

2. Compute the changes in the Helmholtz free energy and the Gibbs free energy when 10 moles of argon gas in a volume of 25 liters at 300°K are expanded to 250 liters at the same temperature. What is the maximum amount of work these 10 moles of gas can do when they undergo this expansion?

3. Suppose the same 10 moles of argon gas are then heated at constant volume to 400°K; at constant pressure. What are the additional changes in Helmholtz and Gibbs free energies of the gas?

4. The electrical potential of a battery is 1.5 volts. What is the drop in its Gibbs free energy when 0.1 mole of electrons is passed through a circuit driven by this battery? One Faraday, that is, one mole of electrons, corresponds to 96,500 Coulombs of charge.

21. AVAILABILITY AND CRITERIA OF MERIT

Frequently systems undergo processes in which they are at states of high thermodynamic potential – – high temperature or high pressure or high chemical potential or any combination of these – – relative to their surroundings. When such a system returns to equilibrium with its surroundings, it may release its potential freely, merely increasing the entropy of the universe, or it may convert some or all of its thermodynamic potential into heat or useful work. The quantity that measures how much work can be extracted when the system comes to equilibrium with its surroundings is called the **available work** or **availability**, which we shall denote by α. Suppose the surroundings have temperature T_o, pressure p_o and chemical potentials $\mu_{o1}, \mu_{o2}, \ldots, \mu_{os}$. Then the availability is defined as

$$\alpha = E + p_o V - T_o S - \sum_{i=1}^{s} \mu_{oi} n_i . \qquad (1\ 6\ 3)$$

Availability is not a conventional state function because it depends on the condition of the surroundings as well as the state of the system itself. However it has values that depend only on the states of the system and surroundings and not on the paths taken to reach those states. In that sense, availability is a state function of the two-component supersystem consisting of the system its and surroundings.

Availability is important because it gives the maximum work that can be delivered to a work-receiver, other than the surrounding medium. Examples of such work-receivers are an automobile driveshaft, a flywheel, an electric generator and a rechargeable battery. It is important in evaluating this potential work to take out the contribution that corresponds to the work done against the surrounding medium itself because that contribution is ordinarily not recoverable. The most important component of that lost work is usually the work needed to expand and push out any gases the process generates against the pressure of the surrounding atmosphere.

206

The following argument shows that availability does give the desired measure of maximum work. The system can do a small increment of work directly, given by −dW; from this, we subtract the work $p_o dV$ done against the external surroundings that are not part of the work-receiver. In addition, the system can lose heat which could be converted to work. The heat *loss* of the system that can be used to generate work is −dQ. (Since, we envision a system providing work and heat, its temperature T must be higher than the temperature T_o of the surroundings if dQ is negative.) The maximum work $−dW_{max,thermal}$, that can be done by −dQ, the heat transferred from the system, is that heat multiplied by the efficiency of an ideal Carnot engine η, Eq. **(116)**:

$$dW_{max,thermal} = -\eta dQ = -[(T - T_o)/T]dQ$$

$$= T_o(dQ/T) - dQ \ . \qquad (164)$$

Thus the maximum work $(−dW_{max})$ that the system can contribute to a work reservoir accompanying a heat change dQ of the system and a direct contribution dW of work to it, is

$$- dW_{max} = T_o(dQ/T) - dQ - dW - p_o dV \ . \qquad (165)$$

(If chemical change also occurs, the system should reflect the chemical work done against the medium, $\Sigma\mu_{oi}dn_i$, giving

$$- dW_{max} = T_o(dQ/T) - (dQ + dW) - p_o dV + \Sigma_i\mu_{oi}dn_i) \ . (166)$$

Because the process that yields the maximum work is reversible, we can set dQ/T equal to the entropy change dS; moreover dQ + dW is, as always, the change in internal energy dE. Therefore Eq. **(165)** becomes $−dE − p_o dV + T_o dS$, equal to the negative of a change in Eq. **(163)**, so

$$d\alpha = dE + p_o dV - T_o dS \qquad (167)$$

(with the addition of $-\Sigma\mu_{oi}dn_i$ if chemical change occurs). Thus the loss in availability is precisely the maximum work that the out-of-equilibrium system could supply to a work receiver when the system returns to equilibrium with its surroundings.

For example, when a hot ingot cools, the principal change in the system is in E; the entropy changes also, but the volume changes almost not at all. This system could do work by giving heat to a heat engine, with a change of availability of approximately

$$\Delta A \approx (E_o - E) - T_o(S_o - S) .$$

The energy change is approximately $C_V(T_o - T)$ and the entropy change is approximately $C_V \ln(T_o/T)$, typically much less than the change of energy. The heat capacity of a metal is roughly 6 cal/mole-deg, as the discussion of Chapter 7 showed. If T = 900°K and T_o = 300°K, then the energy change is approximately −3600 cal/mole and the entropy term gives − 6 $\ln$ (300/900) or 1977 cal/mole, so one mole of metal in the ingot loses availability of 1623 cal. This, therefore, is the maximum work that can be extracted from the heat of the ingot when it cools to the temperature of its surroundings.

In practice, systems are more and more frequently designed to capture and use the heat and pressure that materials would otherwise dissipate to their environment. However designing to make efficient use of those sources of energy is not yet a general practice. Two of the stimuli to increase the capture of heat and pressure are an increasing scarcity of fuel and energy, and an increasing realization of the magnitudes that could be realized by converting availability into actual work.

CRITERIA OF MERIT

Efficiency η was introduced into thermodynamics to give a quantitative expression of how well energy can be converted from heat to work. Carnot's analysis of the ideal two-reservoir heat engine gave an absolute character to the efficiency as a criterion of merit for such an engine because it showed that the ideal Carnot engine had the highest possible efficiency for a two-reservoir engine operating between two fixed reservoir temperatures. Real engines can only do worse.

Efficiency is a useful criterion for many kinds of devices and processes, such as an electric power generating station. The definition of efficiency can be extended to make it more widely applicable if the denominator is generalized to be the total input energy from a source, particularly from a source for which we must pay. This generalization has been called the "first-law efficiency."[1] Another important and traditional criterion of performance is essentially the inverse of the first-law efficiency. It is used to describe cooling systems and is called the **coefficient of refrigerator performance**, equal to the heat taken out of the cold reservoir (the system being refrigerated) divided by the work put into the refrigerator. This criterion is used for heat pumps and air conditioners as well as for refrigerators.

Still another related criterion is the (heat) **coefficient of performance** of heat-powered heating systems and absorption refrigerators, which is given by the ratio of the heat delivered to the system being heated or the heat extracted from the system being cooled, divided by the heat (rather than the work) taken in from the supply reservoir. The criteria defined this way have maximum values that depend solely on temperatures of the system's heat reservoirs. Examples are given in Table 3, and can easily be evaluated, as the formulas indicate.

Efficiency, in its original or generalized sense, is only one kind of criterion of merit. Either way, it measures how well one form of energy is converted into another. In any device or process the Carnot efficiency is an upper limit because it is based on a reversible process. Another kind of criterion of merit is based on the comparison of the maximum work system can do when it is constrained

---

1 "Efficient Use of Energy: A Report of the Summer Study on Technical Aspects of Efficient Energy Utilization" (American Physical Society, New York, 1975.)

TABLE 3

Generalized Criteria of Performance*

System	Numerator	Denominator	Maximum Performance	Name of Criterion
Heat engine operating between T_1 (high) & T_2 (low)	Mechanical work out	Heat in from reservoir at T_1	$1 - T_2/T_1$	Efficiency
Electric motor	Mechanical work out	Electrical work in	1	First law efficiency
Battery	Electrical work out	Chemical potential energy stored	1	First law efficiency
Heat pump driven by electrical motor operating between T_1 and T_2	Heat extracted at T_2	Work done by motor	$1/[1-(T_1/T_2)]$	Coefficient of refrigerator performance
Heat-driven heater, in ambient surroundings at T_0	Heat put into warm reservoir at T_2	Heat from hot reservoir (furnished at T_1)	$\dfrac{1-(T_0/T_1)}{1-(T_0/T_2)}$	Heat coefficient of performance
Absorption refrigerator in ambient surroundings	Heat removed from cool reservoir at T_2	Heat from hot reservoir at T_1	$\dfrac{1-(T_0/T_1)}{(T_0/T_2)-1}$	Heat coefficient of performance

* Based largely on Table 2.1 of "Efficient Use of Energy", loc. cit.

T_1 is the temperature of the hot reservoir, T_2 is the temperature of the cold reservoir, and T_0 is the ambient temperature outside the system (which may be the same as T_2 for some systems).

in realistic ways, with the work that an ideal less constrained system (such as a reversible system) would do. Some such realistic constraints are the inclusion of friction, heat leaks and realistic heat conductances between systems and their reservoirs. A closely related constraint is the condition that the system operate at a finite non-zero rate. Without this constraint, many processes would in principle achieve their ideal limits of efficiency by being operated as reversible, infinitely slow processes. Even processes with certain kinds of imperfections – – specifically, imperfections such as friction, that grow with speed of operation – – approach ideality in the infinitely slow limit. If a designer were to maximize efficiency without including a constraint on rate, the process design would become that of the infinitely slow, reversible limit. This is not the case for systems with energy losses, such as heat leaks, that persist even in the infinitely slow limit of the process. For such systems, the most efficient operation occurs at some finite rate. If heat leaks out of a system, then generally the most efficient rate is achieved when the process operates at a non-zero rate, i.e., in finite time; the faster the process, the smaller the losses from heat leaks, However, the Second Law condition dS > dQ/T tends to push the optimum to longer times. When the balance is reached, an optimal fraction of the heat from the hot reservoir goes into work.

One specific criterion of merit of the second kind, which is closely related to the efficiency and maximum work, can be derived for finite-rate machines. This is the comparison of the *actual power* with the *maximum power*, or energy per unit time, that can be developed. The simplest example of this is the Carnot engine generating work at a finite rate, and connected to its reservoirs through an insulating layer. The maximum power that such an engine can develop was found by Curzon and Ahlborn.[2] If the Carnot engine is connected with reservoirs at T_H and T_L through an imperfect heat conductor, the heat flows at a rate

---

2 F.L. Curzon and B. Ahlborn, *American Journal of Physics*, **43**, 22 (1975); B. Andresen, R.S. Berry, A. Nitzan and P. Salamon. *Physical Review A* **15**, 2086 (1977).

$$dQ/dt = \kappa \Delta T ,$$ (161)

where κ is the heat conductance and ΔT is ($T_{reservoir} - T_{system}$), so that if $T_{reservoir} = T_H$, $T_{system} \approx T_1 < T_H$ then dQ/dt is positive, and heat flows from the reservoir *into* the system. Note that if κ is finite, then the system at its hottest temperature T_1 is always cooler than T_H and the system at its coolest temperature T_2 is always warmer than T_L. The problem of maximum power, which we shall not solve in detail, is one of first finding the optimal T_1 and T_2, from the general expression for the total power

$$\Pi_0 = \text{work per cycle/time per cycle}$$

Maximizing Π_0 with respect to T_1 and T_2 gives

$$T_1 = \tfrac{1}{2}(T_H + \sqrt{T_H T_L})$$ (169)

and
$$T_2 = \tfrac{1}{2}(T_L + \sqrt{T_H T_L}) ,$$ (170)

the system temperatures that maximize the power. The efficiency of the system when it operates to deliver maximum power is

$$\eta_{power} = 1 - \sqrt{T_L/T_H} = 1 - T_2/T_1 < 1 - T_L/T_H .$$ (171)

The maximum power itself is

$$\Pi_{0,max} = \kappa/\gamma(\sqrt{T_H} - \sqrt{T_L})^2$$ (172)

where $\gamma = C_p/C_V$. The optimal period depends on specific properties, notably heat capacities, of the gas and on the mechanical design of the engine, as well as on the heat conductance and reservoir temperatures. The maximum power depends only on the heat conductance, the reservoir temperatures and the heat capacity ratio $C_p/C_V = \gamma$, *not* on mechanical design, a remarkable conclusion. The optimum temperatures T_1 and T_2 and the efficiency require even less information; they depend only on the reservoir temperatures. Note that Eq. **(171)** shows that the efficiency of the engine optimized for power is less than the maximum possible efficiency. We pay a price for operating at a nonzero rate.

Curzon and Ahlborn compared the performance of real electric generating stations of three kinds with the ideal efficiencies for the temperatures of the three systems, and with the maximum theoretical power levels. The results are given in Table 4. Clearly, real electrical power stations come closer to maximizing power than to maximizing efficiency.

Table 4
(Taken from Curzon and Ahlborn, ref 2)

Power source	T_L (°C)	T_H (°C)	Maximum efficiency η (Carnot)	Efficiency at maximum power Π	η (obser ved)
West Thurrock (UK)[a] Coal Fired Steam Plant	ca 25	565	64.1%	40%	36%
CANDU (Canada)[b] PHW Nuclear Reactor	ca 25	300	48.0 %	28%	30%
Larderello (Italy)[c] Geothermal Steam Plant	80	250	32.3%	17.5%	16%

a. Taken from D.B. Spalding and E.H. Cole, *Engineering Thermodynamics* (Edward Arnold, London, 1966) 2nd Ed., pp. 200-286, esp. table, p. 209.

b. Taken from G.M. Griffiths, *Phys. Can.* **30**, 2 (1974), esp. p. 5.

c. Taken from A. Chierici, *Planning a Geothermal Power Plant: Technical and Economic Principles* (U.N. Conference on New Sources of Energy, New York, 1964), Vol 3, pp. 299-311.

The concept of maximizing power output, or of finding the rate of operation that maximizes efficiency, has been part of detailed engineering design in many areas. However its use as a general principle for thermodynamics is surprisingly new, and has hardly been exploited. It is likely to be one of the more important practical directions for the discipline in future years.

One other class of criteria of merit is extremely important. These are the criteria that compare real and ideal performance explicitly. The most frequent forms are ratios of real to ideal performance. The **effectiveness** ε is such a quantity, specifically defined as

$$\varepsilon = \text{(actual work delivered)/(loss in availability)}, \quad \textbf{(173)}$$

which is immediately recognizable as the ratio of the actual work delivered to the maximum work deliverable by the system's coming to equilibrium with its surroundings. A generalization of effectiveness was made[3] as

$$\varepsilon = \frac{\text{(desired heat or work transferred by a device or system)}}{\text{(maximum possible heat or work transferable by the same device or process using the same input as the real system)}} \quad \textbf{(174)}$$

and was given the name "second-law efficiency". An equivalent statement phrased in terms of real and ideal *inputs* to produce a fixed amount of output is

$$\varepsilon = \frac{\text{(minimum availability required to produce the fixed output)}}{\text{(actual availability used to produce the same output)}} \quad .$$

Another related criterion is the waste factor w, which was first defined[4] in terms of a fixed output, rather than in terms of

---

3 "Efficient Use of Energy", loc. cit.
4 R.S. Berry and M.F. Fels, *Bulletin of the Atomic Scientists* (December, 1973), p. 11.

output from a fixed input, like the second statement of generalized effectiveness. The waste factor and the effectiveness are closely related:

$$w = \frac{(\text{real work required} - \text{ideal limit of work required})}{(\text{real work required})}$$

$$= 1 - \frac{w_{ideal}}{w_{real}} \text{ (for fixed output)}$$

$$= 1 - \varepsilon \ . \tag{175}$$

Generalized effectiveness is a powerful guide for identifying processes or devices that offer high returns from technological improvement. Here are some examples.

a) Electric power generation from the combustion of fossil fuels: the work put into electric energy is the numerator and the heat energy from combustion is the denominator, so the effectiveness ε is close in value to the efficiency η. [The same is true for any turbine system, but not for individual stages of a real turbine because in each stage, the work done is less than the actual decrease of availability, which is itself less than the maximum (constant-entropy) decrease of availability, and the latter is equal to the enthalpy or heat which is the denominator of the efficiency.] Typical values given by Keenan put ε at about 80% when η is about 70%.

b) Furnaces fired by fossil fuel: the ideal is complete conversion of the heat of combustion into warmth in the desired space. The availability actually delivered by the fuel is approximately the fuel's heat of combustion, $A \approx Q_c$. Then minimum heat or availability required to do the warming is the heat of combustion Q_c, multiplied by the ratio $(1 - T_{outside}/T_{room})$. The two temperatures are often very close. Suppose $T_{outside}$ = 273°K (32°F) and T_{room} = 294°K (70°F), and that the system has a perfect furnace, which loses no heat of combustion through direct transfer to the outside (equivalent to incomplete heat transfer to the air and walls of the space being warmed.) Then ε = 1 − 273/294, or 0.072. If, as in real furnaces, only 50-80% of the heat of combustion is transferred, the generalized effectiveness is correspondingly reduced.

c)   Preparation of pig iron by blast furnace operation: if we consider the initial and final material states as ore (and other inputs) and pig iron (and other outputs) at ambient temperature and normal atmospheric pressure, thus neglecting the availability of the hot intermediate state, we find that $w = (\Delta g_{real} - \Delta g_{ideal})/\Delta g_{real}$. The Δg's correspond to the transformations actually performed and the ideal thermodynamic limits for a process based on the conversion of Fe_2O_3 and carbon to Fe and CO_2, and of the simultaneous conversion of limestone to lime and CO_2, as it occurs in the blast furnace. The typical value of w is 0.9, so ε is about 0.1. This is a moderately high effectiveness in the primary metals area.

d)   Smelting of copper from copper sulfide: the same method used for the iron blast furnace can be applied to recovery of copper by the reaction of CuS with air to give Cu and SO_2. For this system, w is typically 1.04, and ε, −0.04, which means that energy is put into the process in practice when it should, in ideality, deliver energy. The effectiveness analysis gives us particularly dramatic information in this instance: we are using a potential fuel as an energy sink. There is no basic reason in the scheme of human use of resources why the process should not have an effectiveness below zero, unless there is some driving force to conserve energy and use availability to the utmost. If we perceive natural resources, fuels included, to be increasingly scarce, then we may well have reason to search for potential fuels like copper ore, and develop the technology to permit them to supply energy instead of requiring it.[5]

The analysis of how well systems actually perform, compared with how well they might perform, is only a step in a natural sequence of studies. Thermodynamics was developed to help people make better steam engines, not merely to tell how well real and ideal engines use fuel. Thermodynamics as it is now offers us guidance toward more efficient use of fuels and other natural resources. In time, it will doubtless provide still more precise guides than it does at present, for example by setting criteria for processes operating with even more realistic constraints than those of finite rates, heat leaks, friction and thermal conductivity. Meanwhile, we can use generalized efficiencies and effectivenesses to compare alternative processes and to tell us where new research and development has the *potential* to produce significant improvements in the way we use resources.

---

5   These are drawn from *Efficient Use of Energy*, loc. cit.;
J.H. Keenan, *Thermodynamics* (MIT, Cambridge, Mass., 1970);
E.P. Gyftopoulos, L.J. Lazerides and T.F. Widmer, *Potential Fuel effectiveness in Industry* (Ballinger, Cambridge, Mass., 1974);
and the calculations of Berry and Fels, Ref 4.

Problems

1. Compute the change in availability when 100 moles of a gas (assumed ideal) initially at 500°K and 10 atm pressure goes to a state at 300°K and 1 atm pressure in equilibrium with its surroundings.

2. Under what circumstances is availability the same as the Gibbs free energy? What is the difference between the physical meanings of the Gibbs free energy and the availability?

3. Consider a refrigerator, driven by an engine, and operating between room temperature T_1 and a lower temperature T_2. Show that the coefficient of refrigerator performance is $T_2/(T_1 - T_2)$ or $(\eta^{-1} - 1)$, where η is the efficiency of an ideal Carnot engine operating between T_1 and T_2. What range of values can the coefficient of refrigerator performance assume?

4. Compute the maximum coefficients of refrigerator performance for a refrigerator operating at 280°K in a room at 300°K, and at 250°K in the same room.

5. Derive the expression below for the power Π_0 delivered by the Carnot cycle with finite heat conductance, using Eqs. **(169)** and **(170)** and the assumptions that a) the adiabatic steps occur instantaneously, and b) the system is at constant temperature T_1 and T_2 when it is in contact with the reservoirs at temperature T_H and T_L respectively.

$$\Pi_0 = \kappa \frac{(T_1 - T_2)(T_1 - T_H)(T_2 - T_L)}{T_L T_H - T_1 T_L}$$

6. Graph the effectiveness of a home furnace whose efficiency is 0.7, as a function of the inside room temperature, if the outside temperature is 0°C (273°K).

INDEX

224